SO-AXJ-682

LONE PINE

Perennials

for

British
Columbia

Alison Beck
Marianne Binetti

The Publisher: Lone Pine Publishing

10145 – 81 Ave.	202A, 1110 Seymour St.	1901 Raymond Ave. SW,
Edmonton, AB T6E 1W9	Vancouver, BC V6B 3N3	Suite C, Renton, WA 98055
Canada	Canada	USA

Website: http://www.lonepinepublishing.com

Canadian Cataloguing in Publication Data

Beck, Alison
　Perennials for British Columbia

　Includes index.
　ISBN 1-55105-258-X

　1. Perennials--British Columbia. I. Binetti, Marianne, 1956- II. Title.
SB434.B418 2000　　635.9'32'09711　　C00-910041-5

Editorial Director: Nancy Foulds
Project Editor: Erin McCloskey
Editorial: Erin McCloskey, Eli McLaren
Technical Review: Marian Vaughan
Production Manager: Jody Reekie
Book Design: Heather Markham
Cover Design: Rob Weidemann
Layout & Production: Heather Markham, Elliot Engley
Cartography: Rob Weidemann
Scanning, Separations & Film: Elite Lithographers Co. Ltd.

Photography: all photographs are by Tim Matheson, except David McDonald 6b, 6l, 6u, 7j, 10, 20b, 34a, 43a, 109, 110, 140, 160, 172, 185, 141, 161b, 163a, 169b, 200b, 203b, 230, 233, 251b, 291b; Elliot Engley 41, 42; Horticultural Photography 6f, 7v, 8p, 8r, 8x, 9d, 43b, 123b, 232, 234, 282, 283, 284a, 288, 289b, 293a, 311, 321; Joy Spur 6t, 7e, 7p, 7q, 7u, 9h, 9k, 12, 19, 14, 15, 20a, 108b, 125, 126, 161a, 180, 184, 189, 203a, 215, 217, 218, 231, 310, 312, 320, 328, 329, 335; Peter Thompstone 6v, 6x, 7w, 8h, 9i, 16b, 50, 158, 159b, 167b, 171, 179b, 183, 199, 201b, 216, 236, 238, 239, 251c, 266, 267, 298, 299b, 330, 331a, 334; Therese D'Monte 6q, 6w, 8q, 9a, 155a, 162, 163b, 168, 240, 241, 245, 250, 290, 291a, 313; Valleybrook Gardens 8n, 16a, 111b, 154, 155b, 235, 284b, 285, 289a

Cover Photographs (from left to right) by Tim Matheson, bleeding heart, Himalyan blue poppy, cranesbill geranium, daylily, black-eyed Susan; *by Therese D'Monte,* Michaelmas daisy

We acknowledge the financial support of the Government of Canada through the Book Publishing Industry Development Program (BPIDP) for our publishing activities.

PC: 4

Canadä

Contents

Acknowledgments

We would like to express our appreciation to all who were involved in the making of this project. Special thanks are extended to the following organizations: in Vancouver, BC, to Acadia Community Garden, UBC, Compost Demonstration Garden, Maple Leaf Nurseries, Murray Nurseries, Queen Elizabeth Park, Southlands Nursery, Southside Perennials, Stanley Park, UBC Botanical Garden, Van Dusen Gardens, West Van Florist; in Victoria, BC, to Butchart Gardens; in Rosedale to Minter Gardens; and in Mount Vernon, WA, to Etera Perennials. Additional thanks goes to Peter Thompstone for his generous contribution and involvement in preparing this book.

Pictorial Index

Alphabetical Order, by Common Name

Agapanthus
p. 62

Ajuga
p. 66

Artemisia
p. 70

Astilbe
p. 74

Baby's Breath
p. 78

Balloon Flower
p. 82

Basket-of-gold
p. 84

Bear's Breeches
p. 88

Bee Balm
p. 92

Bellflower
p. 96

Bergenia
p. 100

Black-eyed Susan
p. 104

Bleeding Heart, p. 106   Blue Star, p. 110   Candytuft, p. 112   Cardinal Flower, p. 114

Catmint, p. 118   Cinquefoil, p. 122   Clematis, p. 124   Columbine, p. 128

Coral Bells, p. 132   Coreopsis, p. 134   Cornflower, p. 138   Corydalis, p. 140

Cranesbill Geranium, p. 142   Cushion Spurge, p. 144   Daylily, p. 146   Delphinium, p. 150

Dwarf Plumbago, p. 154   English Daisy, p. 156   Evening Primrose, p. 158   False Solomon's Seal, p. 160

Fleabane, p. 162   Foamflower, p. 166   Forget-me-not, p. 168   Foxglove, p. 170

Gayfeather, p. 174   Gentian, p. 176   Geum, p. 178   Goat's Beard, p. 180

Golden Marguerite, p. 184   Hens and Chicks, p. 186   Himalayan Poppy, p. 188   Hollyhock, p. 190

Hosta, p. 194   Iris, p. 198   Jacob's Ladder, p. 202   Japanese Anemone, p. 204

Japanese Rush, p. 206   Lady's Mantle, p. 208   Lamium, p. 210   Lewisia, p. 214

Lily-of-the-valley, p. 216   Lungwort, p. 220   Lupine, p. 222   Mallow, p. 226

Marsh Marigold, p. 230   Meadow Rue, p. 232   Meadowsweet, p. 236   Michaelmas Daisy, p. 240

Monkshood, p. 244   Mullein, p. 248   Oriental Poppy, p. 250   Pearly Everlasting, p. 252

Penstemon, p. 254   Peony, p. 256   Phlox, p. 260   Pinks, p. 264

Primrose, p. 268   Purple Coneflower, p. 272   Red-hot Poker, p. 274   Rock Cress, p. 276

Rose Campion, p. 280   Rose-mallow, p. 282   Russian Sage, p. 286   Sandwort, p. 288

Saxifrage, p. 290   Scabiosa, p. 292   Sea Holly, p. 294   Sea Pink, p. 298

Sedum, p. 300   Snow-in-summer, p. 304   Spike Speedwell, p. 308   Stonecress, p. 310

Sweet Rocket, p. 312   Sweet Woodruff, p. 314   Thyme, p. 316   Toadflax, p. 320

Trillium, p. 322   Verbena, p. 324   Vinca, p. 326   Violet, p. 328

Wall Rockcress, p. 330   White Gaura, p. 332   Wild Ginger, p. 334   Yarrow, p. 336

Introduction

Perennials are plants that take three or more years to complete their life cycle; lacking a woody stem structure, they normally die back to the ground over winter. These two requirements distinguish them from annuals and biennials and subshrubs and shrubs. However, perennials, such as columbine, peach-leaved bellflower and Jacob's ladder, naturally live for only a scant three years, although their seedlings and offsets keep their presence alive for years to come. Perennials such as bergenia and pinks are fully evergreen. Other perennials that maintain a woody structure in mild climates die back to the ground in colder winters. Adding to the confusion, many favorite denizens of the 'perennial' garden—thyme and candytuft, for example—are in fact subshrubs. Others, such as sweet rocket and foxglove, are actually biennials. But what all of these plants have in common is their reliable presence from one year to the next, either in themselves or their progeny.

It is useful to know what is to expect of a plant: whether it goes dormant in winter or summer; and whether it blooms in spring or summer or whether blooming is based strictly upon temperature, regardless of time. All plants have adapted over time to particular sets of circumstances in which they will grow best.

British Columbia is the most zonally diverse region of Canada, containing regions representing all nine main temperature zones defined by Agriculture Canada. Owing to this diversity, several factors require consideration. For example, 5° F (-15° C) is very different with snow cover or without; in soggy soil or in dry; following a hot summer or a long, cold, wet one. These factors will have more influence on the survival of plants than will temperature. The province has several different types of growing regions, each with a range of temperature that will serve as an indicator of relative hardiness. Recognize first the type of climate in which you garden to better judge hardiness.

The first region is the near mythic west coast rainforest—extending along the cool, rainy, west-facing mountainsides and along the entire length of Vancouver Island. This region is technically the warmest; consequently it is the most hospitable to plants that retain their foliage all year, as well as winter-flowering plants. A challenge here is dealing with excessive rainfall resulting in poor drainage, sour and soggy soil and root rot in otherwise hardy plants. Also, cool summers or warm falls can cause difficulties. Regardless, this area is spectacularly successful for growing perennials. Foliage plants such as hosta, ajuga and lady's mantle, and shade plants like bleeding heart, corydalis, anemones and primrose thrive. Many old favorites, like black-eyed Susan and monkshood, are joined by more esoteric artemisia and cushion spurge, which are only hardy in this area.

The second region is the drier interior, which exists in pockets throughout the province. The interior has a hotter, drier summer, a more clearly defined spring and fall and a colder winter with a more reliable snowcover. As was the case in the rainforest, the assets of the region are also the problems—cold, heat and drought. This area excels at growing the traditional cottage garden perennials—the daisy (e.g., coneflower and tickseed) as well as other delights including delphinium, lupine, iris and yarrow. Peonies are at their best in this region, as are many plants that require pronounced winter chill.

The arbutus region is essentially alpine with cool mountian areas and a warmer range that is the natural home of the arbutus tree. This narrow strip along the coast has thin soil and little rainfall and is the delight of the alpine gardener because the otherwise difficult gentian, lewisia and bellflower thrive here with comparatively little special provision. There are also areas of bog, near desert and true desert, but all of these places have particular specialties along with their challenges.

Inadequate conditions can be adapted to meet plant requirements. Soil can be made lighter or heavier; exposure to wind can be increased or decreased by careful plant placement; soil fertility or pH can be altered; inadequate rainfall can be supplemented by irrigation; excess rainfall can be partially compensated for through improved drainage. Plants that are tender for a region can be protected with branches or mulch, or in extreme cases, taken indoors.

The B.C. gardener is supported by an active and hospitable gardening population with numerous garden clubs throughout the province. Wholesale growers open their gardens to the public and are an invaluable source of ideas and inspiration, as are the various public gardens. The B.C. Nursery Trades Association now publishes a list of members throughout the province. But the chief resources of any gardener are imagination and enthusiasm.

Regional Temperature Map

Zone information not available.

PRINCE RUPERT

PRINCE GEORGE

KAMLOOPS

KELOWNA

VANCOUVER

VICTORIA

AVERAGE ANNUAL MINIMUM TEMPERATURE

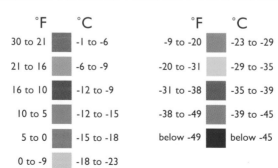

°F	°C		°F	°C
30 to 21	-1 to -6		-9 to -20	-23 to -29
21 to 16	-6 to -9		-20 to -31	-29 to -35
16 to 10	-12 to -9		-31 to -38	-35 to -39
10 to 5	-12 to -15		-38 to -49	-39 to -45
5 to 0	-15 to -18		below -49	below -45
0 to -9	-18 to -23			

Perennial Gardens

Agood perennial garden will be interesting throughout the entire year. Consider the foliage of the perennials you want to use. Foliage can be bold or flimsy, coarse or refined; it can be big or small, light or dark; its color can vary from yellow, gray, blue or purple to any multitude of greens; and it can be striped, splashed, edged, dotted or mottled. The texture can be shiny, fuzzy, silky, rough or smooth. The famous white gardens at Sissinghurst, England, were designed, not to showcase a haphazard collection of white flowers, but to remove the distraction of color and allow the eye to linger on the foliage to appreciate its subtle appeal. Flowers come and go, but a garden planned with careful attention to foliage will always be interesting.

Perennials can be used alone in a garden or combined with other plants such as trees, shrubs and annuals. Perennials form a bridge in the garden between the permanent structure provided by trees and shrubs and the temporary color provided by annuals. They often flower for longer and grow to mature size more quickly that shrubs do and in many cases require less care and are less prone to pests and diseases than annuals.

Perennials can be included in any type, size or style of garden. From the riot of color in a cottage garden or the cool, soothing shades of green in a woodland garden, to a welcoming cluster of pots on a front doorstep, perennials open up a world of design possibilities for even the most inexperienced gardener.

It is very important when planning your garden to decide what you like. If you enjoy the plants that are in your garden then you are more likely to take proper care of them. Decide on what style of garden you like as well as what plants you like. Think about the gardens you have most admired in your neighborhood, in books or while visiting friends. Use these ideas as starting points for planning your own garden.

Select perennials that flower at different times in order to have some part of your garden flowering all season. (See Quick Reference Chart, p. 340.)

Next, consider the size and shape of the perennials you choose. Pick a variety of forms to make your garden more interesting. The size of your garden influences these decisions, but do not limit a small garden to small perennials or a large garden to large perennials. Use a balanced combination of plant sizes that are in scale with their specific location. (See Quick Reference Chart, p. 340.)

There are many colors of perennials. Not only do flower colors vary, but foliage colors vary as well. Different colors have different effects on our senses. Cool colors like blue, purple and green are soothing and make small spaces seem bigger. Warm colors like red, orange and yellow are more stimulating and appear to fill large spaces. (See Quick Reference Chart, p. 340.)

Textures can also create a sense of space. Larger leaves are considered coarse in texture and their visibility from a greater distance make spaces seem smaller and more shaded. Small leaves, or those that are finely divided, are considered fine in texture and create a sense of greater space and light. Some gardens are designed solely by texture.

Variety of textures

Rose-mallow

Bleeding Heart

English Daisy

Coarse-textured Perennials
Bear's Breeches
Bergenia
Black-eyed Susan
Daylily
Hosta
Lungwort
Mullein
Purple Coneflower
Rose-mallow
Sedum 'Autumn Joy'

Fine-textured Perennials
Artemisia
Astilbe
Baby's Breath
Bleeding Heart
Columbine
Coreopsis
Lady's Mantle
Meadow Rue
Thyme

Decide how much time you will have to devote to your garden. With good planning and advanced preparation you can enjoy low-maintenance perennial gardens. Consider using plants that perform well with little maintenance and ones that are generally pest and disease free.

Low-maintenance Perennials
Ajuga*
Bee Balm*
Coral Bells
Cornflower*
Cinquefoil
Daylily*
English Daisy*
Foxglove*
Hosta
Lamium*
Pinks
Scabiosa
(*may take over garden)

Getting Started

Once you have some ideas about what you want in your garden, consider the growing conditions. Plants grown in ideal conditions, or conditions as close to ideal as you can get them, are healthier and less prone to pest and disease problems than plants growing in stressful conditions.

Do not attempt to make your garden match the growing conditions of the plants you like. Choose plants to match your garden conditions. The levels of light, the type of soil and the amount of exposure in your garden provide guidelines that make plant selection easier. A sketch of your garden, drawn on graph paper, may help you organize the various considera-

tions you want to keep in mind as you plan. Start with the garden as it exists. Knowing your growing conditions can prevent costly mistakes—plan ahead rather than correct later.

Sun or Shade

Buildings, trees, fences and even the time of day all influence the amount of light that gets into your garden. There are four basic categories of light in the garden: full sun, partial sun, light shade and full shade. Knowing what light is available in your garden helps in deciding where to put each plant.

Full sun locations receive direct sunlight all or most of the day. An example would be a location along a south-facing wall. Partial sun, or partial shade, locations receive direct sun for part of the day and shade for the rest. An east- or west-facing wall gets only partial sun. Light shade locations receive shade most or all of the day, but some sun gets through to ground level. The ground under a small-leaved tree is often lightly shaded. Small dapples of sun are visible on the ground beneath the tree. Full shade locations receive no direct sunlight. The north side of a house is considered to be in full shade.

It is important to remember that full sun is more intense in some areas than others. On the coast, many ferns, bleeding hearts and hostas will thrive in full sun, but inland these same plants will wilt in anything but partial to full shade. Additionally, the shady side of a building that provides protection from the heat of summer, may also cause a harder and longer freeze than some woodland plants can tolerate.

Perennials for Full Sun

Artemisia
Basket-of-gold
Coreopsis
Daylily
Mallow
Phlox
Russian Sage
Sedum
Thyme

Perennials for Full Shade

Astilbe
Bleeding Heart
Hosta
Lamium
Lily-of-the-valley
Lungwort
Monkshood
Primrose
Saxifrage
Sweet Woodruff
Trillium
Vinca
Violet
Wild Ginger

Sunny, late summer garden

Shade border

Soil

Plants and the soil they grow in have a unique relationship. Many plant functions go on underground. Soil holds air, water, nutrients and organic matter. Plant roots depend upon these resources while using the soil to hold themselves upright.

Soil is made up of particles of different sizes. Sand particles are the largest. Water drains quickly out of sandy soil and nutrients are quickly washed away. Sand has lots of air space and doesn't compact easily. Clay particles are the smallest and can only be seen through a microscope. Water penetrates clay very slowly and drains very slowly. Clay holds the most nutrients, but there is very little room for air and clay compacts quite easily. Most soil is made up of a combination of different particle sizes. These soils are called loams.

Perennials for Sandy Soil

Artemesia
Basket-of-gold
Cornflower
Cinquefoil
Cushion Spurge
Mullein
Rose Campion
Russian Sage
Thyme

Perennials for Clay Soil

Ajuga
Black-eyed Susan
Cranesbill Geranium
Foamflower
Hosta
Lily-of-the-valley
Vinca
Yarrow

Perennials for Woodland Soil

Bleeding Heart
Columbine
False Solomon's Seal
Foamflower
Jacob's Ladder
Lungwort
Primrose
Vinca
Violet
Wild Ginger

The other aspect of soil to consider is the pH. This is the scale on which the acidity or alkalinity is analyzed. Most soils in the Pacific Northwest are acidic. You can test your soil if you plan to amend it. Testing kits are available at most garden centers. Soil acidity influences which nutrients are available for plants. Soil can be made more alkaline with the addition of horticultural lime. If a plant you want to grow likes a very alkaline soil, you might choose to grow it in

Acidic, Pacific Northwest garden

Dry, exposed garden

Moist, woodland garden

a container or raised bed where it will be easier to control and amend the pH as needed. Most plants prefer soil pH between 5.5 and 7.5.

Another thing to consider is how quickly the water drains out of your soil. Rocky soil on a hillside will probably drain very quickly and should be reserved for those plants that prefer a very well-drained soil. Low-lying areas tend to retain water longer and some areas may rarely drain at all. Moist areas can be used for plants that require a consistent water supply and the areas that stay wet can be used for plants that prefer boggy conditions. Drainage can be improved in very wet areas by adding sand or gravel to the soil or by building raised beds. Water retention in sandy soil can be improved through the addition of organic matter.

Perennials for Moist Soil

Astilbe

Bleeding Heart

Blue Star

Cardinal Flower

Goat's Beard

Hosta

Iris

Japanese Rush

Lady's Mantle

Lungwort

Marsh Marigold

Meadowsweet

Monkshood

Primrose

Rose-mallow

Water Forget-me-not

Perennials for Dry Soil
Artemisia
Baby's Breath
Basket-of-gold
Cinquefoil
Coreopsis
Dwarf Plumbago
Evening Primrose
Fleabane
Lupine
Mullein
Pinks
Rock Cress
Russian Sage
Sea Pink
Sedum
Toadflax
Yarrow

Finally, consider the exposure in your garden. Wind, heat, cold and rain are the elements your garden is exposed to, and different plants are better adapted than others to the potential damage these forces can cause. Buildings, walls, fences, hills, hedges and trees can all influence your garden's exposure.

Wind in particular can cause extensive damage to your plants. Plants can become dehydrated in windy locations because they may not be able to draw water out of the soil fast enough to replace the water that is lost through the leaves. Tall, stiff-stemmed perennials can be knocked over or broken by strong winds. Some plants that do not require staking in a sheltered location may need to be staked in a more exposed one. Use plants that are recommended for exposed locations or temper the effect of the wind with a hedge or some trees. A solid wall will create turbulence on the leeward side, while a looser structure, like a hedge, will break up the wind and protect a larger area.

Perennials for Exposed Locations
Basket-of-gold
Candytuft
Columbine
Creeping Phlox
Cushion Spurge
Penstemon
Sedum (groundcover species)
Thyme

Map out your garden's various growing conditions such as shaded areas, wet areas, windy areas and so on. This guideline will help you recognize where your plants will do best.

Dry garden

Preparing the Garden

Taking the time before you start planting to properly prepare the flowerbeds will save you time later on. Removing all weeds and amending the soil with organic matter prior to planting is the first step in caring for your perennials. Thoroughly digging over a bed and picking out all the weeds by hand is the best technique.

Composting

All soils, from the heaviest clay to the lightest sand, benefit from the addition of organic matter. Some of the best additives are compost, well-rotted manure and composted hemlock bark

mulch because they add nutrients as well as improving the soil. These additives improve heavy clay soils by loosening them and allowing air and water to penetrate. Organic matter improves sandy or light soils by increasing the ability of the soils to retain water, which allows plants to absorb nutrients before they are leached away. Mix organic matter into the soil with a garden fork. Within a few months, earthworms will break down the organic matter and at the same time, their activities will keep the soil from compacting.

Compost worms

Weeding

In forests, meadows or other natural environments, organic debris such as leaves and various plant bits break down on the soil surface and the nutrients are gradually made available to the plants that are growing there. In the home garden, where pests and diseases may be a problem and where untidy debris isn't practical, a compost pile or bin is useful. Compost is a great regular additive for your perennial garden and good composting methods will help reduce pest and disease problems.

Compost can be made in a pile, in a wooden box or in a purchased composter, and the process is not complicated. A pile of kitchen scraps, grass clippings and fall leaves will eventually break down if simply left alone. The process can be sped up if a few simple guidelines are followed.

Use dry as well as fresh materials with a higher proportion of dry matter than fresh green matter. Dry matter may be chopped straw, shredded leaves or sawdust, whereas green matter may be vegetable scraps, grass clippings or pulled weeds. The green matter breaks down quickly and produces nitrogen, which composting organisms use to break down dry matter. Spread the green materials evenly throughout the pile by layering them between dry materials.

Composting materials

Wooden compost bins

Plastic compost bins

Layers of soil or finished compost will introduce the organisms that are necessary to break down the compost pile properly. Fertilizers available at garden centers can help speed up the composting process. If the pile seems very dry you can add a bit of water as you layer. The pile needs to be moist but not soggy.

Turn the pile over or poke holes in it with a pitch fork every week or two. Air must get into the pile and will speed up decomposition. A well-aerated compost pile will generate a high degree of heat. A thermometer attached to a long probe, like a meat thermometer, will be able to take the temperature near the middle of the pile. Compost can easily

Hot compost decomposing

reach temperatures of 160° F (71° C). At this temperature weed seeds have been destroyed and many damaging soil organisms killed. Most beneficial organisms are not killed unless the temperature rises above this degree. Once your pile reaches 160° F (71° C) leave it sit. Once you notice the temperature dropping significantly, turn the pile to aerate it and stimulate the process again.

Your compost has reached the end of its cycle when you can no longer recognize the matter that went into it and the temperature no longer rises when you turn the pile. It may take as little as one month to reach this stage and be ready to spread onto your perennial garden. It will have a good mixture of nutrients and be rich in beneficial organisms.

You may want to avoid putting diseased, pest-ridden material into your compost pile. By adding this material, you may risk spreading problems throughout your entire garden. If you do put diseased material in the pile, put it as near the center as possible where the temperatures are highest.

Plant Selection

Plants can be purchased or started from seed. Purchased plants may begin flowering the same year they are planted, while plants started from seed may take several years to mature. Starting plants from seed is more economical if you want large numbers of plants. (See how to start seeds in the Propagation section, p. 40.)

Plants and seeds are available from many sources. Garden centers, mail order catalogs and even friends and neighbors are excellent sources of perennials. Be sure that you get your perennials from a reputable source and that the plants are not diseased or pest-ridden.

As well as garden centers, there are a number of garden societies that pro-mote the exchange of seeds, and many public gardens sell seeds of exclusively rare plants. Gardening clubs are also a great source of rare and unusual plants.

Purchased perennials come in two main forms. They are sold in pots or they are sold bare-root, usually packed in moist peat moss or saw-dust. Potted perennials are growing and have probably been raised in the pot. Bare-root perennials are typi-cally dormant, although some of the previous year's growth may be evi-dent or there may be new growth starting. Sometimes the piece of root appears to have no evident growth, past or present. Both potted and bare-root perennials are good purchases and in each case there are things to look for to make sure that you are getting a plant of the best quality.

Potted plants come in many sizes and though a larger plant may appear more mature it may be better to choose a smaller one that will suffer less from the shock of being transplanted. Most perennials grow quickly once they are planted in the garden. Select plants that seem to be a good size for the pot they are in. When tapped lightly out of the pot, the roots should be visible but not winding and twisting around the inside of the pot. The leaves should be a healthy color.

If the leaves appear to be chewed or damaged, check carefully for insects or diseases before you purchase the plant. If you find insects on the plant you may not want to purchase it unless you are willing to cope with the hitchhikers you are taking home. If the plants are diseased, do not purchase them. Deal with any pest problems before you move the plants into the garden to avoid spreading the pest.

Once you get your plants home, water them if they are dry and keep them in a lightly shaded location until you plant them. Remove any damaged growth and discard it. Try to plant your new perennials into the garden as soon as possible.

Bare-root plants are most commonly sold through mail order, but some are available in garden centers, usually in the spring. Choose roots that are dormant (without top growth). The plant may take longer to establish itself if it is growing before being placed in the garden. It may have too little energy to recover after trying to grow in the stressful conditions of a plastic bag.

Cut off any damaged parts of the roots with a very sharp knife. Bare-root perennials need to be planted more quickly than potted plants because they will dehydrate quickly out of soil. Soak the roots in lukewarm water and either plant them directly in the garden or into pots with good quality potting soil until they can be moved to the garden.

It is often difficult to distinguish the top from the bottom of some bare-root plants. Usually there is a tell-tale dip or stub from which the plant previously grew. If you can't find any distinguishing characteristics, lay the root in the ground on its side and the plant will send the roots down and the shoots up.

Plant on right (not root bound) is healthier   Root-bound plant

Planting Perennials

Once you have your garden planned, the soil well prepared and the perennials ready, it is time to plant. If your perennials have identification tags, be sure to poke them into the soil next to the newly planted perennials. Next spring, when most of your perennial bed is nothing but a few stubs of green, the tags will help you with identification and location.

Potted Perennials

Perennials in pots are convenient because you can space them out across the bed or rearrange them without having to dig. Once you have the collection organized you can begin planting. Do not unpot the plant until immediately before you transplant it or the roots will dry out.

To plant potted perennials, start by digging a hole about the width and depth of the pot. Remove the perennial from the pot. If the pot is small enough, you can hold your hand across the top of the pot letting your fingers straddle the stem of the plant, and then turn it upside-down. Never pull on the stem or leaves to get a plant out of a pot. It is better to cut a difficult pot off rather than risk damaging the plant. Tease a few roots out of the soil ball to get the plant growing in the right direction. If the roots have become densely wound around the inside of the pot, you

Support the plant as you pull off the pot.

should cut into the root mass with a sharp knife to encourage new growth into the surrounding soil. The process of cutting into the bottom half of the root ball and spreading the two halves of the mass outward like butterfly wings is called butter-flying the roots and is a very effective way to promote fast growth of pot-bound perennials that are being transplanted. Place the plant into the prepared hole. It should be buried to the same level that it was at in the pot, or a little higher, to allow for the soil to settle. If the plant is too low in the ground it may rot when rain col-lects around the crown. Fill the soil in around the roots and firm it down. Water the plant well as soon as you have planted it, and regularly until it has established itself.

Bare-root Perennials
During planting, bare-root peren-nials should not be spaced out across the bed unless they are already in pots. Roots dry out very quickly if you leave them lying about waiting to be planted. If you want to visualize your spacing, you can poke sticks into the ground or put rocks down to represent the locations of your perennials.

If you have been keeping your bare-root perennials in potting soil, you may find that the roots have not grown enough to knit the soil together and that all the soil falls away from the root when you remove it from the pot. Don't be concerned. Simply follow the regular root-planting instructions. If the soil does hold together, plant the root the way you would a potted perennial.

Root Types
The type of hole you need to dig will depend on the type of roots the perennial has. Plants with **fibrous roots** will need a mound of soil in

Loosen the root ball before firming it into the ground.

the center of the planting hole to spread the butterflied roots out evenly. The hole should be dug as deep as the longest roots. Mound the soil into the center of the hole up to ground level. Spread the roots out around the mound and cover them with loosened soil. If you are adding just one or two plants and do not want to prepare an entire bed, dig a hole twice as wide and deep as the root ball and amend the soil with composted manure mixed with peatmoss. Add a slow release organic fertilizer to the backfill of soil that you spread around the plant. Fresh chicken or barnyard manure can also be used to improve small areas, but it should be placed in the bottom of the planting hole. Add a layer of soil on top of the manure before you add the plant. Roots that come in contact with the fresh manure will suffer fertilizer burn.

Plants with a **tap root** need a hole that is narrow and about as deep as the root is long. The job is easily done with the help of a trowel: open up a suitable hole, tuck the root into it and fill it in again with the soil around it. If you can't tell which end is up, plant the root on its side.

Some plants have roots that appear to be tap roots, but the plant seems to be growing off the side of the root, rather than upwards from one end. These roots are called **rhizomes**. Iris roots are rhizomes. Rhizomes should be planted horizontal in a shallow hole and barely covered with soil.

In most cases, you should try to get the crown at or just above soil level and loosen the surrounding soil in the planting hole. Keep the roots thoroughly watered until the plants are well established.

Mixed perennials in a planter

Lady's Mantle

Whether the plants are potted or bare-root it is good to leave them alone and let them recover from the stress of planting. In the first month, you will need only to water the plant regularly, weed it and watch for pests. A mulch spread on the bed around your plants will keep in moisture and control weeds.

If you have prepared your beds properly you probably won't have to fertilize in the first year. If you do want to fertilize, and then wait until your new plants have started healthy new growth, apply only a weak fertilizer to avoid damaging the new root growth.

Perennials can also be grown in planters for portable displays that can be moved about the garden. They can be used on patios or decks, in gardens with very poor soil or in yards where kids and dogs might destroy a traditional perennial bed. Many perennials such as hosta and daylily can grow in the same container without any fresh potting soil for five or six years. Be sure to fertilize and water perennials in planters more often than those growing in the ground. Dig your finger deep into the soil around the perennial to make sure it needs water. Too much water in the planter causes root rot.

When designing a planter garden, you can either keep one type of perennial in each planter and display many planters together, or mix different perennials in large planters along with annuals and bulbs. The latter choice results in a dynamic bouquet of flowers and foliage. Keep the tall upright perennials such as yarrow in the

Sedum, catmint and annuals in a planter

Perennials for Planters

Bellflower
Candytuft
Catmint
Cinquefoil
Coreopsis
Cranesbill Geranium
Daylily
Hosta
Lady's Mantle
Penstemon
Pinks
Scabiosa
Saxifrage
Sedum
Verbena
Veronica
Yarrow

center of the planter, the rounded or bushy types like coreopsis around the sides and low growing or draping perennials such as the species candytuft along the edge of the planter.

Perennials that have long bloom times or attractive foliage are good for planters. Remember that planters are more exposed during the winter months. To keep more tender perennials from freezing, move them to a protected location. Choose only the hardiest perennials if you intend to leave the planters outdoors all winter.

Mixed display of perennials and annuals in a wheelbarrow

Care of Perennials

Many perennials require little care, but all will benefit from a few maintenance basics. Weeding, pruning, deadheading and staking are just a few of the chores that, when done on a regular basis, keep major work to a minimum.

Weeding

Controlling weeds is one of the most important things you will have to do in your garden. Weeds compete with your perennials for light, nutrients and space. Weeds can also harbor pests and diseases. Try to prevent weeds from germinating. If they do germinate, pull them out while they are still small and before they have a chance to flower, set seed and start a whole new generation of problems.

Weeds can be pulled out by hand or with a hoe. Quickly scuffing across the soil surface with the hoe will pull out small weeds and sever larger ones from their roots. A layer of mulch is an excellent way to suppress weeds.

Mulching

Mulches are an important gardening tool. They prevent weed seeds from germinating by blocking out the light. Soil temperatures remain more consistent and more moisture is retained under a layer of mulch. Mulch also prevents soil erosion during heavy rain or strong winds. Organic mulches can consist of compost, bark chips, shredded leaves or grass clippings. Organic mulches are desirable because they improve the soil and add nutrients as they break down.

In spring, spread a couple inches of mulch over your perennial beds around your plants. Keep the area immediately around the crown or stem of your plants clear. Mulch that is too close to your plants can trap moisture and prevent good air circulation. If the layer of mulch disappears into the soil over summer, you should replenish it.

A fresh layer of mulch, up to 4" (10 cm) thick, can be laid once the ground freezes in fall, to protect the plants over winter. You can cover the plants with dry material like chopped straw, pine needles or shredded leaves. This covering is especially important in regions where the winters are cold. If your ground doesn't freeze, you can still apply a layer of mulch, but don't cover the crowns of your perennials.

In late winter or early spring, once the weather starts to warm up, pull the mulch layer off the plants and see if they have started growing. If they have, you can pull the mulch back, but keep it nearby in case you need to put it back on to protect the tender new growth from a late frost. Once your plants are well on their way and you are no longer worried about frost, you can remove the protective mulch completely. Compost the old mulch and apply a new spring and summer mulch.

The soil in this garden is protected by mulch.

Deadheading

Deadheading—the removal of flowers once they are finished blooming—serves several purposes. It keeps plants looking tidy, prevents the plant from spreading seeds (and therefore seedlings) throughout the garden, prolongs blooming and helps prevent pest and disease problems.

Deadheading is not necessary for every plant. There are some plants with seedheads that are left in place to provide interest in the garden over winter. Other plants are short-lived and by leaving some of the seedheads in place you are encouraging future generations to replace the old plants. Hollyhock is one example of a short-lived perennial that reseeds. The self-sown seedlings of some perennials often do not possess the attractive features of the parent plant.

Meadowsweet

Perennials with Interesting Seedheads

Astilbe
Clematis
False Solomon's Seal
Goat's Beard
Meadowsweet
Oriental Poppy
Purple Coneflower
Russian Sage
Sedum 'Autumn Joy'

Deadheading asters

Clematis

Flowers can be pinched off or dead-headed by hand or snipped off with hand pruners. Bushy plants that have many tiny flowers, particularly ones that have a short bloom period like basket-of-gold can be more aggressively pruned back with garden shears once they are done flowering. In some cases, e.g., creeping phlox, shearing will promote new growth and blooms later in the season.

Mallow

Perennials that Self-seed

Ajuga
Bleeding Heart (variable seedlings)
Blue Star (variable seedlings)
Cardinal Flower
Forget-me-not
Hollyhock (variable seedlings)
Lady's Mantle
Lupine
Mallow
Mullein
Pinks
Rose Campion
Stonecress (variable seedlings)
Sweet Rocket
Violet

Rock Cress

Perennials to Shear Back After Blooming

Amsonia
Baby's Breath
Basket-of-gold
Candytuft
Creeping Phlox
Cornflower
Bellflower
Blue Star
Cranesbill Geranium
Golden Marguerite
Lamium
Rock Cress
Snow-in-summer
Stonecress
Sweet Woodruff
Thyme
Yarrow

Ajuga and Lady's Mantle

Pruning

Many perennials will benefit from a bit of grooming. Resilient health, plentiful blooming and more compact growth are the signs of a well-groomed garden. Pinching, thinning and disbudding plants before they flower will enhance the beauty of a perennial garden. The methods for pruning are simple but some experimentation is required in order to get the right effect in your own garden.

Thinning is done to clump-forming perennials like black-eyed Susan or bee balm early in the year when shoots have just emerged. These plants develop a dense clump of stems that allows very little air or light into the center of the plant. Remove half of the shoots when they first emerge. This removal will increase air circulation and prevent diseases such as powdery mildew. The increased light encourage more compact growth and more flowers. Throughout the growing season, thin any growth that is weak, diseased or growing in the wrong direction.

Black-eyed Susan

Trimming or pinching perennials is a simple procedure, but timing it correctly and achieving just the right look can be tricky. Early in the year, before the flower buds have appeared, trim the plant to encourage new side shoots. Remove the tip and some stem of the plant just above a leaf or pair of leaves. This can be done stem by stem, but if you have a lot of plants you can trim off the tops with your hedge shears to one-third of the height you expect the plants to reach. The growth that begins to emerge can be pinched again. Beautiful layered effects can be achieved by staggering the trimming times by a week or two.

Perennials to Prune Early in the Season

Artemisia
Bee Balm
Black-eyed Susan
Catmint
Cornflower
Mallow
Michaelmas Daisy
Pearly Everlasting
Purple Coneflower
Rose-mallow
Sedum 'Autumn Joy'

Purple Coneflower

Give plants enough time to set buds and flower. Continual pinching will encourage very dense growth but also delay flowering. Most spring-flowering plants cannot be pinched back or they will not flower. Early summer or midsummer bloomers should be pinched only once, as early in the season as possible. Late summer and fall bloomers can be pinched several times but should be left alone past June. Don't pinch the plant if flower buds have formed—it may not have enough energy or time left in the year to develop a new set of buds. Experimentation and keeping detailed notes will improve your pinching skills.

Disbudding is the final grooming stage. This is the removal of some flower buds to encourage the remaining ones to produce larger flowers. This technique is popular with peony growers.

Staking, the use of poles or wires to hold plants erect, can often be avoided by astute thinning and pinching, but there are always a few plants that will need a bit of support to look their best in your garden. There are three basic types of stakes to go with the different growth habits that need support. Plants that develop tall spikes like hollyhock, delphinium and sometimes foxglove require each spike to be staked individually. A strong narrow pole such as a bamboo stick can be pushed into the ground early in the year and the spike tied to the stake as it grows. A forked branch can also be used to support single-stem plants.

Many plants, such as peony, get a bit top-heavy as they grow and tend to flop over once they reach a certain height. A wire hoop, sometimes called a peony ring, is the most unobtrusive way to hold up such a plant. When the plant is young, the

Wire hoops popular with peony growers

legs of the peony ring are pushed into the ground around it and as the plant grows up, it is supported by the wire ring. At the same time the bushy growth hides the ring. Wire tomato cages can also be used.

Other plants, like coreopsis, form a floppy tangle of stems. These plants can be given a bit of support with twiggy branches that are inserted into the ground around the young plants that then grow up into the twigs.

Some people consider stakes to be unsightly no matter how hidden they seem to be. There are a couple of things you can do to reduce the need for staking. Don't assume a plant will do better in a richer soil than is recommended. Very rich soil causes many plants to produce weak, leggy growth that is prone to falling over. Also, a plant that likes full sun will be stretched out and leggy if grown in the shade. Plants can give each other some support in the border. Mix in plants that have a more stable struc-

Spiral stakes for individual spikes

ture between the plants that need support. A plant may still fall over slightly, but only as far as its neighbor will allow. Many plants are available in compact varieties that don't require staking.

Astilbe in need of support

Watering

Watering is another basic of perennial care. Many perennials need little supplemental watering if they have been planted in their preferred conditions and are given a moisture retaining mulch. The rule of watering is to water thoroughly and infrequently. When you do water, make sure the water penetrates to at least a few inches.

Fertilizing

If you prepare your beds well and add new compost to them each spring you will not need to add extra fertilizer. If you have a limited amount of compost, you can mix a slow-release fertilizer into the soil around your perennials in the spring. Some plants, e.g., delphinium, are heavy feeders that need additional supplements throughout the growing season.

There are many organic and chemical fertilizers available at garden centers. Be sure to use the recommended quantity because too much fertilizer will do more harm than good. Roots can be burned by fertilizer that is applied in high concentrations. Problems are more likely to be caused by chemical fertilizers because they are more concentrated than organic fertilizers.

Propagation

Learning to propagate your own perennials is an interesting and challenging aspect of gardening that can save you money but that also takes time and space. Seeds, cuttings and divisions are the three methods of increasing your perennial population. There are benefits and problems associated with each method.

Seeds

Starting perennials from seed is a great way to propagate a large number of plants at a relatively low cost. Seeds can be purchased or collected from your own or a friend's perennial garden. There are some limitations to propagating from seed. Some cultivars and varieties don't pass on the desirable traits to their offspring. Other perennials take a very long time to germinate, if they germinate at all, and an even longer time to grow to flowering size. There are also many perennials that grow easily from seed and flower within a year or two of being transplanted in the garden. There are challenges and limitations to starting perennials from seed but the pleasure you will receive when your plants finally begin to flower is worth the work.

Specific propagation information is given for each plant, but there are a few basic rules for starting all seeds. Some seeds can be started directly in the garden but it is easier to control temperature and moisture levels and to provide a sterile environment if you start the seeds indoors. Seeds can be started in pots or, if you need a lot of plants, flats. Use a sterile soil mix intended for starting seeds. The soil will generally need to be kept moist but not soggy. Most seeds germinate in moderately warm temperatures of about 57–70° F (14–21° C).

There are many seed-starting supplies available at garden centers. Some supplies are useful but many are not necessary. Seed-tray dividers are useful. These dividers, often called plug trays, are made of plastic and prevent the roots from tangling with the roots of the other plants and from being disturbed when seedlings are transplanted.

All seedlings are susceptible to a problem called 'damping off,' which is caused by soil-borne fungal organisms. An afflicted seedling looks as though someone has pinched the stem at soil level, causing the plant to topple over. The pinched area blackens and the seedling dies. Sterile soil mix, air circulation and evenly moist soil will help prevent this problem.

Fill your pot or seed tray with the soil mix and firm it down slightly—not too firmly or the soil will not drain. Wet the soil before planting your seeds. They may wash into clumps if the soil is watered after the seeds are planted. Large seeds can be placed individually and spaced out in pots or trays. If you have divided inserts for your trays you can plant one or two seeds per section. Small seeds may have to be sprinkled in a bit more randomly. Fold a sheet of paper in half and place the small seeds in the crease. Gently tapping the underside of the fold will bounce or roll the seeds off the paper in a more controlled manner. Some seeds are so tiny that they look like dust.

Preparing the seed tray

Using folded paper to plant small seeds

Watering seeds with spray bottle

These seeds can be mixed with a small quantity of very fine sand and spread on the soil surface. These tiny seeds may not need to be covered with any more soil. The medium-sized seeds can be lightly covered and the larger seeds can be pressed into the soil and then lightly covered. Do not cover seeds that need to be exposed to light in order to germinate. Water the seeds using a very fine spray if the soil starts to dry out. A hand-held spray bottle will moisten the soil without disturbing the seeds.

Plant only one type of seed in the pot or flat. Each species has a different rate of germination and the germinated seedlings will require different conditions than the seeds that have yet to germinate. To keep the environment moist, you can place pots inside clear plastic bags. Change the bag or turn it inside out once the condensation starts to build up and drip. Plastic bags can be held up with stakes or wires poked in around the edges of the pot. Many seed trays come with

Plastic bag covering seeded pot

clear plastic covers which can be placed over the flats to keep the moisture in. Plastic can be removed once the seeds have germinated.

Seeds generally do not require a lot of light in order to germinate, so pots or trays can be kept in a warm, out of the way place. Once the seeds have germinated they can be placed in a bright location but out of direct sun. Plants should be transplanted to individual pots once they have three or four true leaves. True leaves are the ones that look like the mature leaves. (The first one or two leaves are actually part of the seed.) Plants in plug trays can be left until neighboring leaves start to touch each other. At this point the plants will be competing for light and should be transplanted to individual pots.

Young seedlings do not need to be fertilized. Fertilizer will cause seedlings to produce soft, spindly growth that is susceptible to attack by insects and diseases. The seed itself provides all the nutrition the seedling will need. A fertilizer, diluted to ¼ or ½ strength, can be used once seedlings have four or five true leaves.

Seeds have protection devices that prevent them from germinating when conditions are not favorable or from all germinating at once. Staggered germination periods improve the chances of survival. Many seeds will easily grow as soon as they are planted, but others need to have their defenses lowered before they will germinate. Some seeds also produce poisonous chemicals in the seed coats to deter insects.

Perennials to Start from Seed
Corydalis
Delphinium
Pinks
Forget-me-not
Foxglove
Hollyhock
Lady's Mantle
Lupine

Corydalis

Seeds can be tricked into thinking the conditions are right for sprouting. Some thick-coated seeds can be soaked for a day or two in a glass of water to promote germination. This mimics the end of the dry season and the beginning of the rainy season, which is when the plant would germinate in its natural environment. The water softens the seed coat and in some cases washes away the chemicals that have been preventing germination. Rose-mallow is an example of a plant with seeds that need to be soaked before germinating.

Rose-mallow

Other thick-coated seeds need to have their seed coats scratched to allow moisture to penetrate the seed coat and prompt germination. In nature, birds scratch the seeds with gravel in their craws and acid in their stomachs. Nick the seeds with a knife or gently rub them between two sheets of sand paper. Leave the seeds in a dry place for a day or so after scratching them before planting them. This gives the seeds a chance to get ready for germination before they are exposed to water. Lupine and anemone have seeds that need their thick coats scratched.

Japanese Anenome

Plants from northern climates often have seeds that wait until spring before they germinate. These seeds must be given a period of cold weather, which mimics winter, before they will germinate. One method of cold treatment is to plant the seeds up in a pot or tray and place them in the refrigerator for up to two months. Check the container regularly and don't allow these to dry out. This method is fairly simple, but not very practical if your refrigerator is as crowded as mine. Yarrow, bergenia and primrose have seeds that respond to cold treatment.

A less space-consuming method is to mix the seeds with some moistened sand, peat or sphagnum moss. Place the mix in a sealable sandwich bag and pop it in the refrigerator for up to two months, again being sure the sand or moss doesn't dry out. The seeds can then be planted in the pot or tray. Spread the seeds and the moist sand or moss onto the prepared surface and press it down gently.

If the thought of any dirt at all in the refrigerator is unbearable, don't despair, you are living in one of the ideal climates for providing a natural cold treatment to your seeds. Cool, but not excessively freezing winter temperatures can be used to your advantage if you have unheated spot that is sheltered from the rain, such as a porch, garage or cold frame. As long as the temperature in your chosen location doesn't fluctuate too much over the winter and you remember to check that the soil is kept moist, this strategy works well.

Shade cloth protecting small plants in a cold frame

Plastic dome protecting a patio bed

A cold frame is a wonderful tool for the gardener. It can be used to protect tender or moisture-intolerant plants over the winter, to start vegetable seeds early in the spring, to harden plants off before moving them to the garden, to protect fall-germinating seedlings and young cuttings or divisions and to start seeds that need a cold treatment. This mini-greenhouse structure is built so that ground level on the inside of the cold frame is lower than on the outside. The angled, hinged lid is fitted with glass. The soil around the outside of the cold frame insulates the plants inside. The lid lets light in and collects some heat during the day and prevents rain from damaging tender plants. If the interior gets too hot, the lid can be raised for ventilation.

Cuttings

Cuttings are an excellent way to propagate varieties and cultivars that you really like but that don't come true from seed or don't produce seed at all. Each cutting will grow into a reproduction of the parent plant. Cuttings are taken from the stems of some perennials and the roots of others.

Stem cuttings are generally taken in the spring and early summer. During this time plants go through a flush of

Rooted cuttings

fresh, new growth, either before or after flowering. Avoid taking cuttings from plants that are in flower. Plants that are in flower, or are about to flower, are busy trying to reproduce; plants that are busy growing, by contrast, are already full of the right hormones to promote quick root growth. If you do take cuttings from plants that are flowering be sure to remove the flowers and the buds to divert the plant's energy back into growing.

Large numbers of cuttings don't often result in as many plants. Cuttings need to be kept in a warm humid place to root, which makes them very prone to fungal diseases. Providing proper sanitation and encouraging quick rooting will increase the survival rate of your cuttings.

Catmint

Sedum 'Autumn Joy'

Perennials to Propagate from Stem Cuttings

Artemisia
Basket-of-gold
Bellflower
Bleeding Heart
Candytuft
Catmint
Cinquefoil
Clematis
Coreopsis
Cornflower
Cushion Spurge
Michaelmas Daisy
Mullein
Penstemon
Pinks
Rockcress
Sedum 'Autumn Joy'
Snow-in-summer
Thyme
Veronica
Violet
Yarrow

Removing lower leaves

Dipping in rooting hormone

Debate exists over what the size of cuttings should be. Some gardeners claim that smaller cuttings are more likely to root and root more quickly. Other gardeners claim that larger cuttings develop more roots and become established more quickly once planted in the garden. You may wish to try different sizes to see what works best for you. A small cutting is 1–2" (2–5 cm) long and a large cutting is 4–6" (10–15 cm) long.

Firming the cutting into soil

Size of cuttings can be determined by the number of leaf nodes on the cutting. You will want at least three or four nodes on a cutting. The node is where the leaf joins the stem and it is from here that the new roots will grow. The base of the cutting will be just below a node. Strip the leaves gently from the first and second nodes and plant them below the soil. The new plants will grow from the nodes above the soil. The leaves can be left in place on the cutting above ground. If there is a lot of space between nodes, your cutting will be longer than recommended. Some plants have almost no space at all between nodes. Cut these plants to the recommended length and gently remove the leaves

from the lower half of the cutting. Plants with several nodes close together often root quickly and abundantly.

Always use a sharp, sterile knife to make the cuttings. Cuts should be made straight across the stem. Once you have stripped the leaves you can dip the end of the cutting into a rooting-hormone powder intended for softwood cuttings. Sprinkle the powder onto a piece of paper and dip the cuttings into it. Discard any extra powder left on the paper to prevent the spread of disease. Tap or blow the extra powder off the cutting. Cuttings caked with rooting hormone are more likely to rot rather than root and they do not root any

New top growth

Healthy roots

faster than those that are lightly dusted. Your cuttings are now prepared for planting.

The sooner you plant your cuttings the better. The less water the cuttings lose, the less likely they are to wilt and the more quickly they will root. Cuttings can be planted in a similar manner to seeds. Use sterile soil mix, intended for seeds or cuttings, in pots or trays that can be covered with plastic to keep in the humidity. Other mixes to root the cuttings in are sterilized sand, perlite, vermiculite or a combination of the three. Firm the soil down and moisten it before you start planting. Poke a hole in the surface of the soil with a pencil or similar object, tuck the cutting in and gently firm the soil around it. Make sure the lowest leaves do not touch the soil and that the cuttings are spaced far enough apart that adjoining leaves do not touch each other. Pots can be placed inside plastic bags. Push stakes or wires into the soil around the edge of the pot so that the plastic will be held off the leaves.

The rigid plastic lids that are available for trays may not be high enough to fit over the cuttings in which case you will have to use stakes and a plastic bag to cover the tray.

Keep the cuttings in a warm place, about 65–70° F (18–21° C), in bright indirect light. A couple of holes poked in the bag will allow for some ventilation. Turn the bag inside out when condensation becomes heavy. Keep the soil moist. A hand-held mister will gently moisten the soil without disturbing the cuttings.

Most cuttings will require from one to four weeks to root. After two weeks, give the cutting a gentle tug. You will feel resistance if roots have formed. If the cutting feels as though it can pull out of the soil then gently push it back down and leave it for longer. New growth is also a good sign that your cutting has rooted. Some gardeners simply leave the cuttings alone until they can see roots through the hole in the bottoms of the pots. Uncover the cuttings once they have developed roots.

Apply a foliar feed when the cuttings are showing new leaf growth. Plants quickly absorb nutrients through the leaves therefore you can avoid stressing the newly formed roots. Your local garden center should have foliar feeds and information about applying them. Your hand-held mister can apply foliar feeds.

Once your cuttings are rooted and have had a bit of a chance to establish themselves they can be potted up individually. If you rooted several cuttings in one pot or tray you may find that the roots have tangled together. If gentle pulling doesn't separate them, take the entire clump that is tangled together and try rinsing some of the soil away. This should free the roots enough for you to separate the plants.

Pot the young plants in a sterile potting soil. They can be moved into a sheltered area of the garden or a cold frame and grown in pots until they are large enough to plant in the gar-den. The plants may need some protection over the first winter. Keep them in the cold frame if they are still in pots. Give them an extra layer of mulch if they have been planted out.

Basal cuttings involve removing the new growth from the main clump and rooting it in the same manner as stem cuttings. Many plants send up new shoots or plantlets around their bases. Often, the plantlets will already have a few roots growing. The young plants develop quickly and may even grow to flowering size the first summer. You may have to cut back some of the top growth of the shoot because the tiny developing roots won't be able to support a lot of top growth. Treat these cuttings in the same way you would a stem cutting. Use a sterile knife to cut out the shoot. Sterile soil mix and humid conditions are preferred. Pot plants individually or place them in soft soil in the garden when new growth appears and roots have developed.

Perennials to Start from Basal Cuttings

Ajuga
Bee Balm
Bellflower
Catmint
Cranesbill Geranium
Cushion Spurge
Daylily
Delphinium
Hollyhock
Lamium
Lupine
Phlox
Scabiosa
Sedum
Verbena

Lamium

Root cuttings can also be taken from some plants. Dandelions are best known for this trait: even the smallest piece of root sprouts a new plant. But there are perennials that have this ability as well. The main difference between starting root cuttings and stem cuttings is that the root cuttings must be kept fairly dry because they can rot very easily.

Cuttings can be taken from the fleshy roots of certain perennials that do not propagate well from stem cuttings. These cuttings should be taken in late winter or early spring when the ground is just starting to warm up and the roots are just about to break dormancy. At this time, the roots of the perennials are full of nutrients, which the plants stored the previous summer and fall, and hormones are initiating growth.

Keep the roots slightly moist, but not wet, while you are rooting them and keep track of which end is up. Roots must be planted vertical, not horizontal, on the soil, and roots need to be kept in the orientation they held while previously attached to the parent plant. There are different tricks people use to recognize the top from the bottom of the roots. One method is to cut straight across the tops and diagonally across the bottoms.

You do not want very young or very old roots. Very young roots are usually white and quite soft; very old roots are tough and woody. The roots you should use will be tan in color and still fleshy. To prepare your root, cut out the section you will be using with a sterile knife. Cut the root into pieces that are 1–2" (2.5–5 cm) long. Remove any side roots before planting the sections in pots or planting trays. You can use the same type of soil mix the rhizomes and stem cuttings were started in. Poke the pieces vertically into the soil and leave a tiny bit of the end poking up out of the soil. Remember to keep the pieces the right way up.

Keep the pots or trays in a warm place out of direct sunlight. Avoid overwatering them. They will send up new shoots once they have rooted and can be planted in the same manner as the stem cuttings (see p. 45).

Oriental Poppy

Perennials to Propagate from Root Cuttings
Baby's Breath
Bear's Breeches
Black-eyed Susan
Bleeding Heart
Evening Primrose
Japanese Anemone
Mullein
Oriental Poppy
Phlox
Primrose
Sedum

Rhizomes of wild ginger

Rhizomes are the easiest root cuttings with which to propagate plants. Rhizomes are thick, fleshy roots that grow horizontally along the ground, or just under the soil. Periodically, they send up new shoots along the length of the rhizome. In this way the plant spreads. It is easy to take advantage of this feature. Take rhizome cuttings when the plant is growing vigorously (usually in the late spring or early summer).

Dig up a section of rhizome. If you look closely at it you will see that it appears to be growing in sections. The places where these sections join are called nodes. It is from these nodes that feeder roots (smaller stringy roots) extend downwards and new plants sprout upwards. You may even see that small plants are already sprouting. The rhizome should be cut into pieces. Each piece should have at least one of these nodes in it.

Fill a pot or planting tray to about 1" (2.5 cm) from the top of the container with perlite, vermiculite or seeding soil. Moisten the soil and let the excess water drain away. Lay the rhizome pieces flat on the top of the mix and almost cover them with more of the soil mix. If you leave a small bit of the top exposed to the light it will encourage the shoots to sprout. The soil does not have to be kept wet, but you should moisten it when it dries out to avoid having your rhizome rot. Once your cuttings have established themselves they can be potted individually and grow on in the same manner as the stem cuttings (see p. 45).

Perennials to Propagate from Rhizomes
Bellflower
Bergenia
Cornflower
Cranesbill Geranium
Iris
Lily-of-the-valley
Wild Ginger

Dividing a geranium

Divisions

Division is quite possibly the easiest way to propagate perennials. As most perennials grow, they form larger and larger clumps. Dividing this clump once it gets big will rejuvenate the plant, keep its size in check and provide you with more plants. If a plant you really want is expensive, consider buying only one because within a few years you may have more than you can handle.

How often a perennial needs dividing or can be divided will vary. Some perennials, like astilbe, need dividing almost every year to keep them vigorous, while others, like peony, should never be divided, because they dislike having their roots disturbed. Each perennial entry in the book gives recommendations for division. There are several signs that a perennial should be divided:

• the center of the plant has died out
• the plant is no longer flowering as profusely as it did in previous years
• the plant is encroaching on the growing space of other plants sharing the bed.

It is relatively easy to divide perennials. Begin by digging up the entire clump and knocking any large clods of soil away from the root ball. The clump can then be split into several pieces. A small plant with fibrous roots can be torn into sections by hand. A large plant can be pried apart with a pair of garden forks inserted back to back into the clump. Plants with thicker tuberous or rhizomatous roots can be cut into sections with a sharp, sterile knife. In all cases, cut away any

Hosta

Non-Dividing Perennials
Baby's Breath
Balloon Flower
Cushion Spurge
Hosta
Japanese Anemone
Lady's Mantle
Oriental Poppy
Peony
Russian Sage
Trillium

old sections that have died out and replant only the newer, more vigorous sections.

Once your original clump is divided into sections, replant one or two of them into the original location. Take this opportunity to work organic matter into the soil where the perennial was growing before replanting it. The other sections can be moved to new spots in the garden or potted up and given away as gifts to gardening friends and neighbors. Get the sections back into the ground as quickly as possible to prevent the exposed roots from drying out. Plan where you are going to plant your divisions and have the spots prepared before you start digging up. Plant your perennial divisions in pots if you aren't sure where to put them all. Water new transplants thoroughly and keep them well watered until they have re-established themselves.

The larger the sections of the division, the more quickly the plant will re-establish itself and grow to blooming size again. For example, a perennial divided into four sections will bloom sooner than one divided into

ten sections. Very small divisions may benefit from being planted in pots until they are bigger and better able to fend for themselves in the border.

Newly planted divisions will need extra care and attention when they are first planted. They will need regular watering and, for the first few days, shade from direct sunlight. A light covering of burlap or damp newspaper should be sufficient to shelter them for this short period. Divisions that have been planted in pots should be moved to a shaded location.

There is some debate about the best time to divide perennials. Some gardeners prefer to divide perennials while they are dormant, whereas others feel perennials establish themselves more quickly if divided when they are growing vigorously. You may wish to experiment with dividing at different times of the year to see what works best for you. If you do divide perennials while they are growing, you will need to cut back one-third to one-half of the growth so as not to stress the roots while they are repairing the damage done to them.

Problems and Pests

Perennial gardens are both an asset and a liability when it comes to pests and diseases. Perennial beds often contain a mixture of different plant species. Many insects and diseases attack only one species of plant. Mixed beds make it difficult for pests and diseases to find their preferred hosts and establish a population. At the same time, because the plants are in the same spot for many years, the problems can become permanent. The advantage is that the beneficial insects, birds and other pest-devouring organisms can also develop permanent populations.

For many years pest control meant spraying or dusting. The goal was to try to eliminate every pest in the garden. A more moderate approach is advocated today. The goal is now to maintain problems at levels at which only negligible damage is done. Chemicals are the last resort. They cause more harm than good. They endanger the gardener and his or her family and they kill the good organisms as well as the bad ones, leaving the garden open to even worse attacks.

There are four steps in managing pests organically. The cultural controls are the most important. The physical controls come next, followed by the biological controls. The chemical controls should only be used when the first three possibilities have been exhausted.

Cultural controls are the regular gardening techniques you use in the day-to-day care of your garden. Growing perennials in the conditions they prefer and keeping your soil healthy with plenty of organic matter are just two of the cultural controls you can use to keep pests manageable in your garden. Choose resistant varieties of perennials that are not prone to problems. Space perennials so that they have good air circulation around them and are not stressed from competing for light, nutrients and room. Remove plants from the garden if they are constantly decimated by the same pests every year. Remove and destroy diseased foliage and prevent the spread of disease by keeping your gardening tools clean and by tidying up fallen leaves and dead plant matter at the end of the growing season.

Physical controls are generally used to combat insect problems. These include things like picking the insects off the perennials by hand, which is not as daunting a solution as it seems if you catch the problem when it is just beginning. Other physical controls are barriers that stop the insects from getting to the plant or traps that either catch or confuse the insect. The physical control of diseases can generally only be accomplished by removing of the infected perennial parts to prevent the spread of the problem.

Biological controls make use of the populations of natural predators. Animals including birds, snakes, frogs, spiders, lady beetles and certain bacteria can play a role in keeping pest populations at a manageable level. Encourage these creatures to take up permanent residence in your garden. A birdbath and birdfeeder will encourage birds to enjoy your yard and feed on a wide variety of insect pests. Many beneficial insects are probably already living in your garden and they can be encouraged to stay with alternate food sources. Many beneficial insects also eat the nectar from flowers. The flowers of nectar plants like yarrow are popular with many predatory insects.

In a perennial garden you should rarely have to resort to chemicals, but if it does become necessary there are some organic options available. Organic sprays are no less dangerous than chemical ones, but they will break down into harmless compounds because they come from natural sources. The main drawback to using any chemicals is that they may also kill the beneficial insects you have been trying to attract to your garden. Organic chemicals should be available at local garden centers and should be applied at the rates and for the pests recommended on the packages. Proper and early identification of

Bullfrogs eat many insect pests.

problems is vital for finding a quick solution.

Whereas cultural, physical, biological and chemical controls are all possible defenses against insects, disease must be controlled culturally. Prevention is often the only hope: once a plant has been infected, it should probably be destroyed, so as to prevent the spread of the disease.

PESTS

Aphids
Cluster along stems, on buds and on leaves. Tiny; pear-shaped; winged or wingless; green, black, brown, red or gray. Suck sap from plants; cause distorted or stunted growth; sticky honeydew forms on the surfaces and encourages sooty-mold growth. Squish small colonies by hand; brisk water spray dislodges them; many predatory insects and birds feed on them; spray serious infestations with insecticidal soap.

Beetles
Some are beneficial, e.g., lady beetles; others, e.g., June beetles eat plants. Larvae: see borers and grubs. Many types and sizes; usually rounded in shape with hard shell-like outer wings covering membranous inner wings. Leave wide range of chewing damage; cause small or large holes in or around margins of leaves; entire leaf or areas between leaf veins are consumed; may also chew holes in flowers. Pick the beetles off at night and drop them into an old coffee can half filled with soapy water (soap prevents them from floating); spread an old sheet under plants and shake off beetles to collect and dispose of them.

Borers
Larvae of some moths and beetles burrow into plant stems, leaves and/or roots. Worm-like; vary in size and get bigger as they bore through plants. Burrow and weaken stems to cause breakage; leaves will wilt; may see tunnels in leaves, stems or roots; rhizomes may be hollowed out entirely or in part; common problem in iris plants, except Siberian Iris, which is resistant to borers. Remove and destroy parts that are being bored; may be able to squish borers within leaves; may need to dig up and destroy infected roots and rhizomes.

Bugs (True Bugs)
Many are beneficial; a few are pests. Small, up to ½" (1 cm) long; green, brown, black or brightly colored and patterned. Pierce plants to suck out sap; toxins may be injected that deform plants; sunken areas are left where pierced; leaves rip as they grow; leaves, buds and new growth may be dwarfed and deformed. Remove debris and weeds from around plants in fall to destroy overwintering sites; pick off by hand and drop into soapy water; spray with insecticidal soap.

Cutworms
Larvae of some moths. About 1" (2.5 cm) long; plump, smooth-skinned caterpillars; curl up when poked or disturbed. Usually only affects young plants and seedlings, which may have been completely consumed or chewed off at ground level where you find them lying in the morning. Create

barriers from old toilet tissue rolls to make collars around plant bases; push tubes at least halfway into ground.

Grubs

Larvae of different beetles; problematic in lawns; may feed on perennial roots. Commonly found below soil level; usually curled in a C–shape; body is white or gray; head may be white, gray, brown or reddish. Eat roots; plant is wilting despite regular watering; entire plant may pull out of the ground with only a gentle tug in severe cases. Toss any grubs you find while digging onto a stone path or patio for birds to find and devour; control populations by applying parasitic nematodes or milky disease spore to infested soil (ask at you local garden center).

Leafminers

Larvae of some flies. Tiny; stubby maggots; yellow or green. Tunnel within leaves leaving winding trails; tunneled areas lighter in color than rest of leaf; unsightly rather than health risk to plants. Remove and destroy infected foliage; remove debris from area in fall to destroy overwintering sites; attract parasitic wasps with nectar plants like yarrow.

Slugs & Snails

Common pest in Northwestern gardens. Slugs lack shells; snails have a spiral shell; slimy, smooth skin; can be up to 8" (20 cm) long, many are smaller; gray, green , black, beige, yellow or spotted. Leave large ragged hole in leaves and silvery slime trails on and around plants. Attach strips of copper to wood around raised beds or smaller boards inserted around sus-ceptible groups of plants; slugs and snails will get shocked if they try to cross copper surfaces; pick them off by hand in the evening; spread wood ash or diatomaceous earth (available in garden centers) on ground around plants, which will pierce their soft bodies and cause them to dehydrate.

Spider Mites

Almost invisible to the naked eye; relatives of spiders without their insect-eating habits. Tiny; eight-legged; may spin webs; red, yellow or green; usually found on undersides of plant leaves. Suck juice out of leaves; may see fine webbing on leaves and stems; may see mites moving on leaf undersides; leaves become discolored and speckled in appearance, then turn brown and shrivel up. Wash them off with a strong spray of water daily until all signs of infestation are gone;

Slug damaged hosta

predatory mites are available through garden centers for ornamental plants or spray plants with insecticidal soap.

Thrips
Difficult to see; may be visible if you disturb them by blowing gently on an infested flower. Tiny; slender; narrow fringed wings; yellow, black or brown. Suck juice out of plant cells, particularly flowers and buds, causing mottled petals and leaves, dying buds and distorted and stunted growth. Remove and destroy infected plant parts; encourage native predatory insects with nectar plants like yarrow; spray severe infestations with insecticidal soap.

Whiteflies
Tiny flying insects that flutter up into the air when the plant is disturbed. Tiny; moth-like; white; live on undersides of plant leaves. Suck juice out of plant leaves, causing yellowed leaves and weakened plants; leave sticky honeydew on leaves encouraging sooty mold growth. Destroy weeds where insects may live; attract native predatory beetles and parasitic wasps with nectar plants like yarrow; spray severe cases with insecticidal soap.

DISEASES

Anthracnose
Fungus. Yellow or brown spots on leaves; sunken lesions and blisters on stems; can kill plant. Choose resistant varieties and cultivars; remove and destroy infected plant parts; thin out stems to improve air circulation; avoid handling wet foliage; keep soil well drained; clean up and destroy material from infected plants at end of growing season.

Aster Yellows
Transmitted by insects called leafhoppers. Stunted or deformed growth; leaves yellowed and deformed; flowers dwarfed and greenish; can kill plant. Control insects with insecticidal soap; remove and destroy infected plants; destroy any local weeds sharing these symptoms.

Botrytis Blight
Fungal disease. Leaves, stems and flowers blacken, rot and die. Remove and destroy any infected plant parts; thin stems to improve air circulation, keep mulch away from the base of the plant, particularly in the spring, when the plant starts to sprout; remove debris from the garden at the end of the growing season.

Leaf Spot
Two common types: one caused by bacteria and the other by fungus. *Bacterial:* small speckled spots grow to encompass entire leaves; brown or purple in color; leaves may drop. *Fungal:* black, brown or yellow spots causing leaves to wither. Controls are similar for both types of leaf spot though the bacterial infection is more severe. Remove and destroy infected plant parts; remove entire plant with bacterial infection and sterilize tools; avoid wetting foliage or touching wet foliage; remove debris at end of growing season.

Mildew
Two types: both are caused by fungus, but they have slightly different symptoms. *Downy mildew:* yellow spots on upper sides of leaves and downy fuzz

on the undersides; fuzz may be yellow, white or gray. *Powdery mildew:* white or gray powdery coating on leaf surfaces that doesn't brush off (photo on p. 54). Choose resistant cultivars; space plants well; thin stems to encourage air circulation; remove and destroy infected leaves or other parts; tidy any debris in fall.

Nematodes

Tiny worm-like organisms that give plants disease symptoms. One type infects foliage and stems the other infects the roots. *Foliar:* yellow spots that turn brown on leaves; leaves shrivel and wither; problem starts on lowest leaves and works up the plant. *Root knot:* plant is stunted; may wilt; yellow spots on leaves; roots have tiny bumps or knots. Remove infected plants; mulch soil; clean up debris in fall; don't touch wet foliage; add organic matter and parasitic nematodes to soil.

Rot

Several different fungi that affect different parts of the plant. *Crown rot:* affects base of plant, causing stems to blacken and fall over and leaves to yellow and wilt; can kill plant. *Root rot:* leaves yellow and plant wilts; digging up plant will show roots rotted away. Keep soil well drained; don't damage plant if you are digging around it; keep mulches away from plant base; destroy infected plants.

Rust

Fungus. Pale spots on upper leaf surfaces; orange, fuzzy or dusty spots on leaf undersides. Destroy infected plant parts; choose rust-resistant varieties and cultivars; avoid handling wet leaves; provide plant with good air circulation; clear up garden debris at end of season.

Sooty Mold

Fungus. Thin black film forms on leaf surfaces that reduces amount of light getting to leaf surfaces. Wipe mold off leaf surfaces; control insects like aphids and whiteflies (honeydew left on leaves forms mold).

Viruses

Plant may be stunted and leaves and flowers distorted, streaked or discolored. Viral diseases in plants cannot be controlled. Destroy infected plants; control insects like aphids, leafhoppers and whiteflies that spread disease.

Wilt

If watering hasn't helped consider these two fungi. *Fusarium wilt:* plant wilts, leaves turn yellow then die; symptoms generally appear first on one part of the plant before spreading to other parts. *Verticillium wilt:* plant wilts; leaves curl up at the edges; leaves turn yellow then drop off; plant may die. Both wilts are difficult to control. Choose resistant varieties and cultivars; clean up debris at end of growing season; destroy infected plants; solarize soil before replanting (this may help if you've lost an entire bed of plants to these fungi).

You can make your own insecticidal soap at home. Mix one teaspoon (5 ml) of mild dish detergent or pure soap (biodegradable options are available) with one quart (1 L) of water in a clean spray bottle. Spray the surface areas of your plants and rinse them well within an hour of spraying.

About this Guide

The perennials in this book are organized alphabetically by their common names. Additional common names and Latin names appear after the primary reference. We chose to use local common names instead of the sometimes less familiar scientific names. Later in the text, we describe our favorite recommended or alternate species, but keep in mind that many more hybrids, cultivars and varieties are often available. Check with your local greenhouses or garden centers when making your selection.

Quick identification information on flower color, height and spread as well as when to expect the plant to bloom are the first details given on each plant. At the back of the book there is a Quick Reference Chart, which is a handy guide to planning the diversity in your garden.

Our region encompasses a broad geographical diversity, ranging from coastal sites to mountain tops. In addition, there is great variation in the changes in the seasons, which occur at different times of the year depending on region. It is important to understand your relative proximity to the coast or mountain ranges and how far north or south you are in order to understand the duration, intensity and corresponding months of your seasons. We have referred to the seasons in the general sense, because spring, for example, can start as early as February, depending on where you garden.

The Perennials

Agapanthus
Lily-of-the-Nile
Agapanthus spp.

Flower color: Purple or blue, sometimes white.
Height: 36–60" (90–150 cm) tall. **Spread:** 24" (60 cm) wide.
Blooms: Summer to early fall.

Globes of starry blue flowers and strap-like leaves give this perennial a tropical look that hints back to its warm climate origins. I love this plant mixed with tropical-looking annuals such as 'Terrace Lime' Sweet Potato Vine or the bright, bold-colored coleus. Around a pool or deck, pots of agapanthus will transform the space with a relaxed, resort look.

Agapanthus is derived from the Greek words agape *meaning 'love' and* anthos *meaning 'flower.'*

PLANTING

Seeding: Seeds may be planted in spring. Maintain soil temperature at 55–59° F (13–15° C). Seedlings will be slow to develop and may not come true to type.

Planting out: Spring. Keep plant well watered in first year to help it become established.

Spacing: 24" (60 cm) apart.

GROWING

Grow agapanthus in a **full sun** location. Soil should be **fertile** and **well drained**. Agapanthus prefers a **moist** soil but is drought-tolerant once well established. Agapanthus is an attractive plant whether or not it is in flower. The strap-like leaves are bright green and the round or pendulous clusters of flowers atop their long straight stems make excellent companions to flowering shrubs. The deep blue varieties looking particularly striking with yellow-flowered companions such as *Potentilla fruticosa*.

Protect the seedlings over their first winter. Grow agapanthus in pots and the plants will be easier to move to a protected location for winter, and you can insure that the roots have perfect drainage. Divide in spring whenever it begins to outgrow its allocated spot or when more plants are desired. Agapanthus should begin to flower after two or three years.

RECOMMENDED

Headbourne hybrids are a group of hybrids developed in England from *A. campanulatus*. They are all hardy and vigorous, bearing strong flower colors. **'Alice Gloucester'** has pure white flowers. **'Bressingham Blue'** has deep blue flowers. **'Loch Hope'** has purple-blue flowers. **'Luly'** has light blue flowers.

A. inapertus is a hardy species with pendulous flowers. This species is least likely to grow towards the sun, owing to its erect habit. Varieties are available in all shades of purple-blue.

GARDENING TIPS

Agapanthus doesn't mind cool winter weather but shouldn't be allowed to freeze. It may be grown in containers that are easy to put in a garage during frosty weather, or it may be lifted from the garden and replanted each spring once the weather remains above freezing. Planting it close to the foundation of the house and mulching thoroughly is also a good way of protecting it through winter.

Another consideration when choosing a location is the amount of standing water in the soil in winter. Though this plant likes a lot of water it doesn't like to sit in it for extended periods. Choose a well-drained area or prepare the planting hole with some added gravel in the bottom. Once again, a container plant will be easier to keep dry during winter while it is dormant, since it may be placed in a rain-free zone on a covered porch or patio.

A south-facing bed is a good idea because the plant has a tendency to grow towards the sun and will face southwards with no regard to the orientation of the bed.

PROBLEMS & PESTS

Slugs and snails may sometimes be problems. Bulb and root rots may occur in poorly drained soils.

Ajuga
Carpet Bugleweed
Ajuga reptans

Flower color: Purple or blue, sometimes pink or white.
Height: 4–10" (10–25 cm) tall. **Spread:** 18" (45 cm) or wider.
Blooms: Late spring to early summer.

Ajuga is a lovely groundcover for all seasons. It is the perfect plant to carpet the ground beneath large trees or to use in young gardens as a filler between other plants. The flowers, though pretty, are rather unobtrusive. The leaves of the plant, depending on the variety, are available in shades of bronze, splashed multicolor green, pink and cream or variegated silver.

Bugleweed is widely used in homeopathic remedies for throat and mouth irritations.

PLANTING

Seeding: Not recommended. Often reverts to green-leaved variety from seed.

Planting out: Anytime of year.

Spacing: 18" (45 cm) apart.

GROWING

Ajuga grows well in **full sun to full shade** conditions. Sunny locations produce the best foliage color. Ajuga grows well in any type of soil. It tolerates moist soil as long as it is well drained. Divide at any time of the growing season.

RECOMMENDED

'**Braunherz'** is the best of the deep bronze varieties. It is compact and richly colored in foliage all year.

'**Burgundy Glow'** has leaves in shades of bronze, burgundy and green.

'**Catlin's Giant'** has larger bronze leaves, short spikes of bright blue flowers.

'**Multicolor'** has bronze leaves splashed with pink and white (photo on opposite page).

'**Variegata'** has leaves that are silver-edged green. This cultivar shows its best color in shady locations.

According to European folk myths, bugleweed causes fires if brought into the house.

ALTERNATE SPECIES

A. genevensis (Geneva Bugleweed) is an upright, non-invasive species. The flowers are in shades of blue, pink or white. This species grows up to 12" (30 cm) tall and spreads no more than 18" (45 cm). It is a good choice for a rock garden where it will not attempt to take over the entire site.

'Burgundy Glow'

GARDENING TIPS

Ajuga is an excellent plant to grow in a difficult location such as an exposed slope or under a shady tree. It is also an attractive groundcover in the shrub border where the dense growth will prevent the spread of all but the most tenacious weeds.

This plant's spread may be somewhat controlled by the use of bed-edging materials. It could easily take over the lawn because it spreads readily by stolons (above-ground shoots) and its low growth form escapes the lawn mower blades. If it starts to take over, Ajuga is easy to rip out and the soil it leaves behind will be soft and loose from the penetrating roots. Use Ajuga as a scout plant to send ahead and prepare the soil before you plant anything too fancy in a shaded or woodland garden.

Any new growth or seedlings that don't show the hybrid leaf coloring should be removed.

PROBLEMS & PESTS

Crown rot is the biggest problem for Ajuga. Don't let it sit in water and be sure there is good air circulation around the plants. The purple-leaved varieties are attractive to slugs.

'Catlin's Giant'

Artemisia
Wormwood; Mugwort; Sagebrush
Artemisia spp.

Flower color: White or yellow.
Height: 3–60" (8–150 cm) tall. **Spread:** 12–24" (30–60 cm) wide.
Blooms: Summer. Flowers are insignificant.

This plant is grown for its foliage not for its flowers. The silvery or gray-green leaves of this mound-forming plant make an excellent contrast with bright or pastel-colored flowers of other plants. It is a great perennial for the carefree garden. You don't have to remember to feed or water artemisia, and the gray leaves do not attract slugs. Plant the taller artemisias in rustic pots with sedums and succulents spilling from the sides, and use the low-growing types spilling over the sides of a rock garden or bordering beds next to deep purple *Heuchera*. On a mound of dry rocky soil, the Purple Smoke Tree (*Cotinus coggygria*) underplanted with gray artemisia is a breathtaking sight.

PLANTING

Seeding: Plant in cold frame in fall or spring.

Planting out: Any time during growing season.

Spacing: 12–24" (30–60 cm) apart.

GROWING

Artemisia likes to be grown in **full sun**. Soil should be on the poor side as an overly rich soil will cause lanky, invasive growth. Artemisia likes to be dry, so a **well-drained** soil is absolutely necessary. This plant is very drought-tolerant.

Use artemisia in the border to create a background for bright or soothing flower colors. It also makes an interesting addition to a foliage bed where the fine textured, often lacy leaves contrast with hostas and ferns. Most species of this plant will need dividing every couple of years as they tend to spread quite rapidly.

A. ludoviciana

There are almost 300 species of Artemisia throughout the world.

There are two possible sources of the genus name: it quite possibly honors the botanist and medical researcher from 353 BC, Artemisia, who was the sister of King Mausolus; the other possibility is that it was named after Artemis also known as Diana, the Goddess of the Hunt and the Moon.

RECOMMENDED

'Powis Castle' has finely cut silver foliage. Do not cut back into the old wood or it could kill the plant. Prune in late spring and shear in summer to keep tidy. It grows 24" (60 cm) or taller and spreads 36" (90 cm).

A. lactiflora (White Mugwort) is the only species with interesting flowers. Sprays of white, fragrant flowers are produced in late summer. White Mugwort grows up to 60" (150 cm) tall and spreads 24" (60 cm) wide. It is important to note that deer absolutely love to eat this species. Prune back hard in fall.

***A. ludoviciana* ssp. *albula* 'Silver King'** (Western Mugwort) is native to western North America. It grows 48" (120 cm) tall and 24" (60 cm) wide and may be invasive.

A. schmidtiana (Silvermound) is a smaller, more compact species. It grows 12" (30 cm) tall and 18" (45 cm) wide and has a habit of flopping over if it grows any taller. Flopping over is more likely to happen in hot climates but trimming back this species will solve the problem. **'Nana'** (Dwarf Silvermound) is an even smaller cultivar of this species. 'Nana' grows only 3" (8 cm) tall and 12" (30 cm) wide. It is often used in rock gardens and to edge borders.

A. stelleriana (Beach Wormwood) is native to eastern North America. It grows up to 6" (15 cm) tall and 12–18" (30–45 cm) wide. It is tolerant of oceanside conditions such as salt spray.

A. schmidtiana 'Nana'

GARDENING TIPS

Artemisia responds well to pruning in late spring. If you prune before May, frost may kill any new growth. Whenever it begins to look straggly it may be cut back hard to encourage new growth and maintain a neater form. Smaller forms may also be used to create knot gardens.

If you want to control horizontal spreading, plant artemisia in a bottomless container. Sunk into the ground the container is hidden and it will prevent the plant from spreading beyond the container's edges. With the bottom missing, good drainage can also be maintained.

PROBLEMS & PESTS

Various leaf and stem afflicting fungi such as powdery or downy mildew and rust are possible. A healthy plant in a bright, well-drained location is unlikely to fall victim to these problems. Slug-resistant.

A. lactiflora

Astilbe
False Spiraea
Astilbe x *arendsii*

Flower color: White, pink, purple, peach or red.
Height: 24–48" (60–120 cm) tall. **Spread:** 18–36" (45–90 cm) wide.
Blooms: Early, mid- or late summer, depending on cultivar.

Astilbes are valued in the garden for both their beautiful divided foliage and their plumes of tiny, colorful flowers that rise above the leaves. The flowers last for about a month in summer after which the seedpods remain providing a natural and interesting focus in late summer and fall gardens.

Astilbes are great as cut flowers and if you leave the plumes in a vase as the water evaporates, you'll have dried flowers to enjoy all winter.

PLANTING

Seeding: Not recommended. Plants do not come true to type.

Planting out: Spring. Transplant ½" (1 cm) deeper than when potted.

Spacing: 18–36" (45–90 cm) apart.

GROWING

Astilbes are **shade-loving** plants that will do best when planted in light or partial shade. Heavy shade will reduce the number of flowers. Soil conditions are very important. A **humus-rich, acidic,** deeply prepared, **moist, well-drained** soil is essential. Astilbes like to have moist soil in summer, but should not sit in standing water during winter while they are dormant. Therefore, a site that is well drained in winter is essential.

Astilbes are excellent plants to grow around the edges of a bog garden or pond. They also make good plants for a woodland garden and may be used as shrub-substitutes in a border. Mulch well to prevent summer moisture loss. Divide every three years in spring or fall or the plants will die out. Give Astilbe high nitrogen plant food such as manure or lawn fertilizer in October.

RECOMMENDED

There are many varieties of Astilbe; the following are just a few of the more popular.

'Avalanche' is a white-flowered, late summer cultivar that grows up to 36" (90 cm) tall.

'Bressingham Beauty' has pink, midsummer blooms and grows up to 36" (90 cm) tall.

'Cattleya' has red-pink, midsummer blooms and grows up to 36" (90 cm) tall.

'Etna' is a red-flowered, early to midsummer cultivar that grows up to 30" (75 cm) tall.

'Fanal' is a red-flowered, early summer bloomer that grows 24" (60 cm) tall.

'Weisse Gloria' is a white-flowered, early summer bloomer that grows 24" (60 cm) tall.

ALTERNATE SPECIES

A. chinensis 'Pumila' is a mat-forming groundcover more tolerant of dry conditions than other species. It is a dwarf species growing to 10" (25 cm) tall with dark pink flowers. Grow it in semi-shade.

A. japonica 'Deutschland' is a clump-forming hybrid with pure white flowers. It grows about 20" (50 cm) tall.

GARDENING TIPS

The root crown of Astilbe tends to lift out of the soil as the plant grows. This problem can be solved by applying a top dressing of rich soil as a mulch when the plant starts lifting or by lifting the entire plant and replanting it deeper into the soil.

The flowers of Astilbe fade to various shades of brown. The flowers may be removed once flowering is finished or they may be left in place. This plant self-seeds easily and the flowerheads look interesting and natural in the garden well into fall.

PROBLEMS & PESTS

A variety of pests can on occasion attack Astilbes. Powdery mildew, bacterial leaf spots and fungal leaf spots are also possible problems.

A. japonica 'Deutschland'

In late summer, transplant seedlings found near the parent plant for plumes of color throughout the garden.

Baby's Breath

Gypsophila paniculata

Flower color: White or pink.
Height: 12–48" (30–120 cm) tall. **Spread:** 12–48" (30–120 cm) wide.
Blooms: Most of summer.

The airy little flowers of this delicate perennial contrast well next to broad-leaved plants. Wet soil and rain are the enemies of this drought-loving perennial, but the low-growing varieties seem to handle the wet weather west of the Cascade Mountains better than the taller types.

*The flowers are excellent
for floral arrangements
and last for a long time
if dried first.*

PLANTING

Seeding: Sow in cold frame in spring. Some varieties will not grow from seed.

Planting out: Spring.

Spacing: 36" (90 cm) apart. Small varieties may be planted close together.

GROWING

Locate plants in **full sun**. Soil has to be **neutral or alkaline, of average fertility and very well drained**. Baby's Breath easily rots in areas with moist, acidic soil.

Baby's Breath ties together other plantings in the border with its cloud-like flower clusters. Baby's Breath will develop a large, thick taproot that should not be disturbed once it is established. There is no need to divide Baby's Breath.

RECOMMENDED

'**Fiesta Pink**' and '**Fiesta White**' are the best of the newer cultivars, bearing large heads of fully double flowers in clear pink and pure white, respectively. Fiesta Pink has a more spreading habit, but both are very hardy and reliably produce excellent crops of flowers.

'**Perfecta**' has white, double flowers. It grows 48" (120 cm) tall and wide.

'**Pink Star**' has pink, double flowers. It grows up to 18" (45 cm) tall.

'**Rosy Veil**' is a mound-forming hybrid with pink double flowers. It grows about 16–20" (40–50 cm) tall and up to 36" (90 cm) wide.

'**Viette's Dwarf**' is a smaller plant with white, double flowers. It grows 12–16" (30–40 cm) tall.

'Rosy Veil'

Baby's Breath should not be divided but can be propagated by root or basal cuttings.

PROBLEMS & PESTS

Baby's Breath attracts slugs. Other problems are crown gall, bacterial soft rot or crown or stem rot.

ALTERNATE SPECIES

G. reptans is a low-growing species. It grows up to 8" (20 cm) tall and spreads to form a mat 12–20" (30–50 cm) wide. This species is more acid tolerant than *G. paniculata* and may be used in rock gardens, on rock walls, on pathway edges or at the front of borders.

GARDENING TIPS

Some cultivars are grafted onto stronger rootstock. When planting these cultivars it is important to plant so that the graft union—a fat knob where the two plants are joined—is about an inch below the soil surface. This will encourage the plant above the graft to grow roots of its own.

In order to prolong the flowering period it is best to deadhead Baby's Breath. This sounds easy enough until the plant is covered in tiny blooms, all in different stages of development. Instead of trying to take off the flowers that have faded, wait until the entire plant is almost finished flowering and shear the entire plant back lightly. This will encourage new growth and a second flush of blooms later in the season.

G. reptans

Balloon Flower

Platycodon grandiflorus

Flower color: Blue, pink or white.
Height: About 24" (60 cm) tall. **Spread:** 12" (30 cm) wide.
Blooms: Summer.

A favorite old-fashioned flower of gardeners. Kids delight in popping the air-filled buds between their fingers. Adults don't have to resist this temptation either—popping the buds will not harm the flowers. Blue is a difficult color to come by in the perennial garden but this plant blooms a true medium blue that looks lovely with yellow, pink and purple flowers. Plant it behind the border of low-growing bushy plants so it can spill forward and dangle its balloon-like buds over the edge.

PLANTING

Seeding: Start in containers or direct sow in garden in spring.

Planting out: Spring.

Spacing: 12" (30 cm) apart.

GROWING

Plant in a **full or partial sun**. Soil should be **average to rich, light, moist and well drained**. Use these flowers in a border or rock garden. Division, though rarely required and not recommended because of the deep roots, may be done in spring or early summer. This plant will bloom in the second year after seeding and is popular in cut flower arrangements.

RECOMMENDED

'Park's Double Blue' has blue-purple, double flowers.

'Shell Pink' has very light pink flowers.

P. g. var. *albus* has blue-veined, white flowers.

GARDENING TIPS

Be very careful when dividing this plant. It does not like having its roots disturbed and may take a long time to re-establish itself. If you are only dividing to propagate the plant you would be better off gently detaching from the main root the new shoots that grow up alongside the parent plant. These plants are also prolific self-seeders and new plants can be obtained this way.

PROBLEMS & PESTS

Slugs and snails may be problems, but the plant may also occasionally get leaf spot or blight.

When using these flowers in arrangements it is advisable to singe the cut ends with a lit match to prevent the milky white sap from running.

Basket-of-Gold

Aurinia saxatilis

Flower color: Yellow or occasionally apricot.
Height: 6–12" (15–30 cm) tall. **Spread:** 12" (30 cm) wide.
Blooms: Early through midspring.

This plant is one of the showiest of spring-blooming perennials. The foliage is completely hidden under a blanket of bright flowers. Quick-growing and self-seeding, Basket-of-gold soon takes over in its corner of the garden. It is attractive when grown in a rock garden or at the top of a retaining wall where it can cascade over the edge. Spoon-shaped leaves remain attractive throughout summer.

PLANTING

Seeding: Sow seeds in containers in cold frame in fall or spring.

Planting out: Early spring should allow blooms in first year.

Spacing: 12–18" (30–45 cm) apart. Do not move or divide established plants.

GROWING

Plant in a **full sun** location. Basket-of-gold is drought-tolerant, preferring an **average to sandy, well-drained** soil, and may rot in wet soil. It will flop over in rich soil. Basket-of-gold is useful at the front of a border, in a rock garden or along the top of a retaining wall. It is also a useful plant to use as a ground cover on a sun-exposed or difficult to mow bank. Basket-of-gold should never be divided, but can be sheared back to rejuvenate the plant. New plants can be obtained from cuttings taken after flowering in early summer.

Basket-of-gold belongs to the Brassica family, which includes such plants as cabbage and broccoli.

RECOMMENDED

'Citrina' has lemon yellow flowers.

'Compacta' has golden yellow flowers and only grows 6" (15 cm) tall.

'Dudley Nevill' has apricot-colored flowers.

'Gold Ball' grows 8" (20 cm) tall in a clump with the bright yellow flowers reaching above.

'Variegata' has lemon yellow flowers and cream-colored, irregular margins on the leaves.

GARDENING TIPS

Shearing Basket-of-gold back lightly after flowering will keep the plant compact and will occasionally encourage a few more flowers. Do not shear off all flowerheads in hot regions, as the plants do not live as long when exposed to high temperatures. Self-seeding will provide new plants once the old ones die.

Avoid planting Basket-of-gold near slow-growing plants because it may quickly choke them out. In areas of cool summers, trim back hard after flowering to keep plant healthy and attractive.

PROBLEMS & PESTS

Slug and pest resistant.

Cuttings can be taken from the new growth that starts after the spent flowerheads are sheared back.

Bear's Breeches

Acanthus mollis

Flower color: White in combination with pink, mauve, purple or yellow.
Height: 60" (150 cm) tall. **Spread:** 36" (90 cm) wide.
Blooms: Late spring to early summer.

Bear's Breeches is a very dramatic, large plant that takes up a lot of room in width and height. The plant looks something like a thistle and many of the related species may rival thistles for prickliness. *A. mollis* is the least spiny species. The flowers are borne on long spikes and are usually bi-colored in combinations of white and another color. The flower spikes may be up to 36" (90 cm) long.

Acanthus is derived from the Greek word akanthos *(a thorn) in reference to the spiny nature of this plant.*

PLANTING

Seeding: Start seeds in containers of moist soil in cold frame in spring.

Planting out: Any time during growing season.

Spacing: 36" (90 cm) apart.

GROWING

Locate this plant in **full sun or partial shade**. Bear's Breeches will grow in just about any **well-drained** soil. It prefers a **rich, moist** soil, but you may wish to plant it in a slightly poorer soil as it can be very aggressive. Be sure to keep the plant well watered if it is planted during hot, dry weather to help it become established. It is drought-tolerant but will do best if given an occasional soaking when the weather is hot and dry. This plant dislikes humid conditions.

Bear's Breeches is bold and dramatic and forms large clumps. It makes an excellent central planting in an island bed or in the background of a border. It has an interesting enough appearance to be grown for the leaves alone. The large flower spikes add a dramatic touch and make interesting additions to both fresh and dried arrangements.

Wear gloves when handling Bear's Breeches to prevent any irritation that may occur from the spines.

This plant may require frequent division. Divisions should be made in fall or early spring. Mulch in fall. This plant is fairly hardy. It may die if temperatures drop below 20° F (-7° C).

Grow this plant in the warmest part of the garden and mulch thoroughly. It can also be grown in large containers that can be moved to a warmer spot in winter. Another option is to pot up fall divisions, keep them sheltered over the winter and plant them into the garden in spring.

ALTERNATE SPECIES

A. spinosus (Spiny Bear's Breeches) is very similar to *A. mollis* except that it is very spiny and more tolerant of humid conditions.

GARDENING TIPS

Invasiveness is the most important thing to keep in mind when dealing with this plant. It spreads by rhizomes (roots traveling underground) and even a small piece left in the ground may start a new plant. It can be kept in check by providing natural barriers or by planting it where it has plenty of space to grow.

PROBLEMS & PESTS

Snails and slugs may become problems. Powdery mildew may cause trouble if the plant has poor ventilation. Bear's Breech is also susceptible to fungal and bacterial leaf spots. Both mildew and leaf spots can be prevented if the plant has adequate ventilation and low humidity.

A. spinosus

Bee Balm
Bergamot
Monarda didyma

Flower color: Red, pink, purple or cream.
Height: 36" (90 cm) or taller. **Spread:** 24" (60 cm) wide.
Blooms: Summer.

The brightly colored starry blooms look like ruffles around the neck of a clown. This flower always appears to be having a good time in the garden—blooming, flopping and sprawling about in a relaxed manner and inviting bees and butterflies to visit often. A great companion for herbs because both can be harvested for culinary use.

This plant is named after the Spanish botanist and physician Nicholas Monardes (1493–1588).

PLANTING

Seeding: Start seeds in cold frame in spring or fall.

Planting out: Spring or fall.

Spacing: 24" (60 cm) apart.

GROWING

Bee Balm will do equally well in **full sun or partial shade**. Soil should be of **average fertility, humus-rich, moist and well drained**. Locate along streams or pondsides or in the dappled shade of a woodland border. Bee balm may be used in a border if adequate water is provided. Divide every two or three years in spring, before the new growth emerges but prune back low to the ground every spring.

RECOMMENDED

'Gardenview Scarlet' has large scarlet red flowers and is resistant to powdery mildew.

'Marshall's Delight' has pink flowers and is resistant to powdery mildew. This cultivar cannot be started from seed.

The common name, Bergamot, comes form the scent of the Italian Bergamot orange (Citrus bergamia) often used in aromatherapy.

'**Panorama**' comes in scarlet, salmon or rose pink.

'**Raspberry Wine**' has red blooms on dark green foliage. This cultivar is an excellent choice for cut flowers.

GARDENING TIPS

Bee Balm will attract bees, butterflies and hummingbirds to your garden. Avoid the use of pesticides that can seriously harm or kill these creatures especially if you plan to ingest this, or any plant, in your garden.

The fresh or dried leaves may be used to make a refreshing, minty, citrus-scented tea. Put a handful of fresh leaves in a teapot, pour boiling water over the leaves and let steep for at least five minutes. Sweeten the tea with honey to suit your own taste.

Powdery mildew can be avoided if good air circulation is provided and the plant isn't allowed to dry out for extended periods. To help prevent powdery mildew, thin the stems in spring. If mildew strikes after flowering, cut the plants back to 6" (15 cm) to increase air circulation.

Poor soil makes for a more fragrant, compact plant. In better soil, *Monarda* is larger, lusher and longer blooming.

PROBLEMS & PESTS

Powdery mildew is the worst problem. Other possible problems are rust and leaf spot. Deer and slug resistant.

Bellflower
Campanula spp.

Flower color: Blue is most common.
Also available in white, pink and light or dark purple.
Height: 4–60" (10–150 cm). **Spread:** 12–60" (30–150 cm).
Blooms: Spring, early summer.

These enchanting old-fashioned flowers were fabled to children as fairy caps. The color is a true sky blue that looks especially nice with yellow and pink. Tall bellflowers are wonderful companions for the ruffled pink, shorter growing blooms of Sweet Williams, while the shorter bellflowers are natural companions for pinks and saxifrages. They can also be enjoyed indoors as cut flowers.

PLANTING

Seeding: Start seeds uncovered in cold frame in spring or fall.

Planting out: Spring or fall.

Spacing: 12–60" (30–150 cm) apart, varies according to species.

GROWING

Campanula will grow in many different conditions from **full sun to full shade** depending on which species it is. They will perform well in any soil that is **fairly fertile and well drained**. There are so many forms of *Campanula* that there is one for almost any location. The spreading and trailing forms look good in a rock garden or spilling over a wall. The upright and mounding types look good in a border or cottage garden. Bellflowers make excellent companions for roses in pink, red or yellow because the colors offset each other nicely. Divide them in spring every third or fourth year or more often if they are becoming invasive.

C. rotundifolia

Bellflower can be propagated by basal, new-growth or rhizome cuttings.

C. parsicifolia

RECOMMENDED

C. carpatica (Carpathian Harebell) is a mounding type with blue, purple or white flowers. It grows up to 12" (30 cm) tall and spreads 12–24" (30–60 cm) wide.

C. lactiflora (Milky Bellflower) is an upright type with white to pale blue flowers, sometimes in shades of darker blue and purple. It grows 48–60" (120–150 cm) tall and spreads 24" (60 cm).

C. persicifolia (Peach-leaved Bellflower) is an upright species and is good for cut flowers. It grows 36" (90 cm) tall and spreads 12" (30 cm).

C. portenschlagiana (Dalmatian Bellflower) is a spreading dwarf type with deep purple flowers. It grows 4–6" (10–15 cm) tall and 20" (50 cm) wide and works well in rock gardens.

C. poscharskyana (Serbian Bellflower) is a trailing type with pale to dark purple flowers. It grows 6" (15 cm) tall and spreads 24" (60 cm) wide.

GARDENING TIPS

Bellflower appreciates summer mulch to keep its roots cool. It is important to divide bellflower every two years in early spring or late summer or the plants will die out. Bellflower will have prolonged blooming if deadheaded. Use scissors to cut back one-third of the plant at a time, allowing other sections to continue blooming. As pruned sections starts to bud, cut back other sections for continued blooming.

PROBLEMS & PESTS

The worst problems are with slugs and snails. Bellflower also sometimes has problems with vine weevils, spider mites, aphids, powdery mildew, rust and fungal leaf spots.

There are over 300 species of Campanula found throughout the Northern Hemisphere in habitats ranging from high rocky crags to boggy meadows.

C. poscharskyana

Bergenia

Bergenia spp.

Flower color: Red, deep rose pink to light pink or white.
Height: 12–24" (30–60 cm) tall. **Spread:** 12–24" (30–60 cm) wide.
Blooms: Early spring.

The huge pink or white trusses welcome spring by blooming with the daffodils, and the large flat leaves contrast with the more delicate foliage and flowers of other rock-garden plants such as Creeping Phlox and Candytuft. Use in a raised bed where this plant can enjoy the perfect drainage and moist soil it craves. It also makes a nice edging under the skirts of mature rhododendrons.

PLANTING

Seeding: Rarely comes true to type but fresh ripe seeds will produce the best results. Sow seeds uncovered with soil temperature at 69–70° F (20–21° C). Plants should be ready to be planted out in about two or three months.

Planting out: Spring.

Spacing: 10–20" (25–50 cm) apart.

GROWING

Plant bergenia in **full sun or partial shade**. Soil should be **average to rich and well drained**. Bergenia prefers a moist soil, but is drought-tolerant once it is established. Bergenia is a versatile, low-growing, spreading plant that may be used as a groundcover, as an edging plant for beds, at borders and along pathways, in rock gardens and for mass plantings under trees or shrubs. Divide in spring, every three to five years, when the clump begins to die out in the middle.

Bergenia makes a nice feature plant in spring, blooming at the same time as tulips and Basket-of-gold. When the flowers are finished the leathery leaves help cover the bare spots left by spring-flowering bulbs, adding bulk and substance to the garden all season.

Bergenia is sometimes referred to as an evergreen perennial. This reference can be misleading because during winter the plant draws moisture from the leaves turning them from green to a reddish color.

RECOMMENDED

'Bressingham White' grows up to 12" (30 cm) tall with pure white flowers.

'Evening Glow' or **'Abendglut'** grows up to 12" (30 cm) tall and 18–24" (45–60 cm) wide. The flowers are a deep magenta-crimson. The leaves turn red and maroon in winter.

'Winter Fairy Tale' grows 12–18" (30–45 cm) tall and 18–24" (45–60 cm) wide. Flowers are a deep rose red. The deep green leaves are touched with red in winter.

B. ciliata (Winter Bergenia) grows 18–24" (45–60 cm) tall with equal width. The flowers are white or light pink.

B. cordifolia (Heart-leaved Bergenia) grows up to 24" (60 cm) tall and 24" (60 cm) or wider. The flowers are shades of deep pink. The leaves of this species produce brilliant colors in fall.

B. x schmidtii grows up to 12" (30 cm) tall and 24" (60 cm) wide. The flowers are pink.

GARDENING TIPS

Once flowering is complete, in early spring, bergenia still makes a beautiful addition to the garden, with its thick, leathery, glossy leaves. Bergenia provides a background for other flowers with its soothing expanse of green. As well, many varieties turn attractive colors of bronze and purple in fall and winter.

Propagating by seed can be somewhat risky. You may not get what you hoped for. For more certainty, propagate your plants with root cuttings. Bergenia spreads just below the surface by rhizomes, which may be cut off in pieces and grown separately as long as a leaf shoot is attached to the section.

PROBLEMS & PESTS

The worst problems for bergenia in the Pacific Northwest will be slugs and snails. Occasional problems with fungal leaf spots and root rot as well as weevils, caterpillars and foliar nematodes are also possible.

Another common name for this plant is elephant ears because of the large leathery leaves.

Black-Eyed Susan

Rudbeckia spp.

Flower color: Yellow, orange or red with centers typically brown or green.
Height: 24–72" (60–180 cm) tall. **Spread:** 18–36" (45–90 cm) wide.
Blooms: Midsummer to fall.

No other plant says 'country garden' like *Rudbeckia*, and a cut bouquet of this perennial in a crock or pitcher makes any room look cozy. Along a split-rail fence or in an island planter-bed are great places to plant black-eyed Susan or anyplace you want a casual look with an easy-to-grow plant.

PLANTING

Seeding: Start seeds in cold frame in early spring.

Planting out: Spring.

Spacing: 24" (60 cm) apart.

Black-eyed Susan is a variable plant with some compact forms growing no more than 12" (30 cm) tall and others growing as high as 10' (3 m).

GROWING

Grow these plants in **full sun or partial shade**. Soil should be of **average fertility, fairly heavy and well drained**. A regular supply of moisture is a good idea, although established plants are fairly drought-tolerant.

These brightly flowered native plants make an excellent addition to wildflower and natural gardens. They may be included in borders for their late-season bloom. The flowers last well when cut for arrangements. Divide in spring or fall, every three to five years.

RECOMMENDED

R. fulgida var. *sullivantii* **'Goldsturm'** grows to be 24–30" (60 cm) tall. It is compact with bright yellow, orange or red, brown-centered flowers. This cultivar is resistant to powdery mildew.

R. lactiniata is a large plant than can grow to well over 60" (150 cm) tall. **'Goldquelle'** is a more refined cultivar growing to only 36" (90 cm) tall and bearing bright yellow, double flowers.

R. nitida **'Herbstsonne'** grows up to 72" (180 cm) tall—with extremely tall, but sturdy stems—and 36" (90 cm) wide. The flowers are bright yellow with green centers. The attractive seedheads will bring birds to fall gardens.

GARDENING TIPS

Black-eyed Susan is a low maintenance, long-lived, tough perennial. It is best planted in drifts. Pinching plants in June will make shorter, bushier stands.

PROBLEMS & PESTS

Slugs and snails may be problems with young plants. There may also be occasional problems with aphids, rust, smut and some leaf spots. Deer resistant.

R. fulgida

Bleeding Heart

Dicentra spp.

Flower color: Pink, white, yellow, purple or red.
Height: 15–48" (38–120 cm) tall. **Spread:** 18–36" (45–90 cm) wide.
Blooms: Spring and summer.

Few plants are as graceful in the garden as bleeding hearts. Arching wands suspend delicate, intricate dangling flowers that, to the romantic gardener, look like broken hearts. Cut a few stems for a bud vase where the details of the flowers can be enjoyed up close. The fern-like foliage and tiny, pendulous flowers make this plant very desirable for woodland or shade gardens.

PLANTING

Seeding: Start in cold frame in spring or summer. Plants self-seed in garden.

Planting out: Spring.

Spacing: 18–36" (45–90 cm) apart.

GROWING

Bleeding hearts prefer a **partially shady** location but will grow in sun or shade. Soil should be **moist and humus-rich.** Very dry conditions will cause the plant to die back during summer. Bleeding hearts are versatile plants that look good in many places in the garden. In a woodland garden they may be left to naturalize. They also look good in a mixed border or when used as early season specimen plants. Division is needed infrequently. Common Bleeding Heart doesn't need to be divided at all, while Western Bleeding Heart may be divided every three years or so after it goes dormant or in fall.

D. spectablis

D. spectabilis

By pulling apart the petals of the inverted flowers you can see fairy boats or dancing thumbelinas.

RECOMMENDED

D. exima (Fringed Bleeding Heart) is 15–24" (38–60 cm) tall, with lacy, fern-like foliage. It is a hardy native of the Pacific Northwest. **'Luxuriant'** has blue-green leaves and cinnamon-scented pink flowers.

D. formosa (Western Bleeding Heart) is a low-growing, wide-spreading Pacific native. Growing to 18" (45 cm) tall and 24–36" (60–90 cm) wide. The flowers are pink and fade to white as they mature. It self-seeds freely. ***D. f.* var. *alba*** has white flowers. This species is the most drought-tolerant one, often continuing to flower all summer despite a lack of water. Can be invasive, spreading by underground roots.

D. spectabilis (Common Bleeding Heart) is a magnificent plant, growing to 48" (120 cm) tall and 18" (45 cm) wide. The flowers are rose pink with white inner petals. This plant likes decidu-ous shade. It will die back in summer unless it is watered regularly.

D. formosa

GARDENING TIPS

Though these plants prefer to remain evenly moist they are quite drought-tolerant, particularly if the weather doesn't get hot. Most species will not die from a lack of water, but will merely go dormant. It is most important for them to remain moist while in flower in order to prolong the flowering period. Constant summer moisture will keep the flowers coming until midsummer.

PROBLEMS & PESTS

Slugs and snails may be problems on new growth. Downy mildew, *Verticillium* wilt, viruses, rust and fungal leaf spot may occasionally be problems. Deer resistant.

These delicate plants are the perfect addition to the moist woodland garden. Plant them next to a shaded pond or stream.

D. exima

Blue Star

Amsonia spp.

Flower color: Light blue.
Height: 12–36" (30–90 cm) tall. **Spread:** 12–36" (30–90 cm) wide.
Blooms: Spring and early summer.

The soft and elegant foliage on this perennial resembles the gentle willow. The light blue flowers complement yellow or pink companions in the perennial border. When fall arrives you'll enjoy a two season bonus from this plant as the crisp, cool evening air turns the foliage yellow, making this a good choice to pair with fall mums and asters.

PLANTING

Seeding: Sow seeds in flats in cold frame in fall.

Planting out: Spring or fall.

Spacing: 12–24" (30–60 cm) apart.

GROWING

Blue star prefers to grow in a **partly shaded** location but will tolerate full sun if it is well watered. It is somewhat drought-tolerant if located in partial shade but prefers a **moist well-drained** soil. It will grow best in **average or rich** soil.

The willow-like foliage of blue star turns an attractive yellow in fall and its love of moist soil makes it a beautiful addition at the side of a stream or pond. After flowering, cut it back by one-third with hedge shears to avoid floppy plants. It can be propagated with softwood (new growth) cuttings. Divide in early fall or spring.

This easy-to-grow and hardy perennial is an excellent woodland plant, growing naturally at the margins between forest and field.

A. tabernaemontana var. salicifolia

RECOMMENDED

A. ciliata (Willow Blue Star) flowers in summer and has narrow downy leaves that cluster with the flowers at the ends of the stems. It is native to the southern United States.

A. orientalis flowers in early to midsummer. It is a Mediterranean native growing to the same size, but more densely than the American natives.

A. tabernaemontana var. salicifolia (Willow Blue Star) flowers in spring and has narrow willow-like foliage that turns yellow and orange in fall. Native to the eastern United States, it is the hardiest *Amsonia* species.

GARDENING TIPS

Be sure to wash your hands thoroughly after handling the plants as some people find the sap irritates their skin.

Plants will reseed by the hundreds in moist soil. Cut back spent blossoms to stop new seedlings.

PROBLEMS & PESTS

May have occasional problems with rust.

Candytuft

Iberis sempervirens

Flower color: White.
Height: About 12" (30 cm) tall. **Spread:** 16" (40 cm) or wider.
Blooms: In spring, for up to two months.

A clump of Candytuft with red tulips emerging from its clouds of white blossoms is a happy way to celebrate spring. Candytuft looks great with spring bulbs, especially rock-garden tulips. It is a long-lasting cut flower for miniature bouquets arranged in fancy perfume bottles or tea cups and is perfect paired with the sweet faces of pansies and other short-stemmed spring flowers.

PLANTING

Seeding: Direct sow in garden in spring or fall.

Planting out: Spring.

Spacing: 6–12" (15–30 cm) apart.

GROWING

Candytuft grows well in **full sun**. Soil should be **poor to average, moist and well drained** and preferably **neutral to alkaline**. Candytuft makes an excellent edging plant. It may be used along borders, in rock gardens, beside paths and in the crevices of a rock wall. This plant looks very good when mixed with spring-blooming bulbs. Division is rarely required.

RECOMMENDED

'**Autumn Snow**' bears white flowers in spring and then again in fall.

'**Little Gem**' produces many flowers in spring and is a very compact plant, growing only 6" (15 cm) tall and 10" (25 cm) wide.

'**Purity**' has larger flowers, a longer blooming time and more compact growth.

GARDENING TIPS

Candytuft should be sheared back by about one-third once it has finished flowering to promote new, compact growth. Every two or three years it should be sheared back by one-half to two-thirds to discourage the development of too much woody growth and to encourage abundant flowering. As the stems spread outwards they may root where they touch the ground. These rooted ends may be cut away from the central plant; one may thus avoid attempting to divide the entire plant. In spring, cut away any brown sections resulting from winter damage.

If you are arriving home after dusk on a spring night, this flower will glow in the moonlight to welcome you.

PROBLEMS & PESTS

The biggest problems are likely to be slugs and snails. Caterpillars, clubroot, damping off, gray mold and a few fungal spots are also occasionally possible.

Cardinal Flower

Lobelia cardinalis

Flower color: Red, pink, white, yellow or blue.
Height: 36" (90 cm) tall. **Spread:** 12" (30 cm) wide.
Blooms: Summer and fall.

The colorful blooms of Cardinal Flower will be admired on warm afternoons in late summer. It may be difficult to grow, but gardeners with boggy soil will appreciate this native flower. Many common cultivars are red, inspiring the reference to the beautiful red bird that is also named Cardinal. This is a good perennial to use at the back of the bed so lower growing plants can help support it.

PLANTING

Seeding: Direct sow in garden or in containers in cold frame.

Planting out: Spring.

Spacing: 12" (30 cm) apart.

GROWING

Cardinal Flower will do equally well in **full sun or partial shade**. Soil should be **fertile, slightly acidic and moist.** This is an ideal plant to grow alongside a stream or pond. It may also be grown in a bog garden, as long as it doesn't dry out for extended periods. Pinch the plants in early summer for more compact growth. Divide in summer lifting the entire plant and removing the new rosettes growing at the base. Replant them immediately in the garden. A light mulch and leaving the stems on the plant will help ensure winter survival. Remove the mulch in spring.

Mathias de l'Obel (1538–1616) was the Flemish botanist that Lobelia *was named after.*

RECOMMENDED

'Ruby Slippers' has deep ruby red flowers.

'Twilight Zone' has light pink flowers.

L. c.* var. *alba has white flowers.

ALTERNATE SPECIES

L. siphilitica (Blue Cardinal Flower) grows 24–48" (60–120 cm) tall and 12" (30 cm) wide. In late summer and lasting through fall it bears tall spikes of bright blue flowers.

L. siphilitica

GARDENING TIPS

This plant may require a more acidic soil than other plants growing along a pond. If this is the case it may be planted in a container of acidic soil and sunk into the ground at the edge of the pond.

Cardinal Flower is likely to self-seed quite easily. As this is a short-lived plant, lasting about four or five years, self-seeding is an easy way to insure continuing generations of plants. If you remove the spent flower spikes be sure to allow at least a few of them to remain to spread their seeds. Don't worry too much though, the lower flowers on a spike may have already set seed before the top flowers are finished opening.

Deadheading may encourage a second set of blooms.

PROBLEMS AND PESTS

Occasional problems with slugs, rust, smut and leaf spots are possible.

Catmint

Nepeta spp.

Flower color: Blue or purple are most common. Occasionally white or yellow.
Height: 18–36" (45–90 cm) tall. **Spread:** 12–24" (30–60 cm) wide.
Blooms: Spring and early summer, with possible reblooming in late summer or fall.

If catmint makes you think of drunken felines draped over your shrubbery, you should know that it is mostly *N. cataria* (Catnip) that puts kitty in the party mood. The other catmints can be real workhorses of the garden bed, offering season-long blooms on sturdy, trouble-free plants. In England, where the climate is similar to our own, catmint is a popular plant for bordering pathways and edging flowerbeds.

PLANTING

Seeding: Start seeds in cold frame in fall. Many popular hybrids and cultivars cannot be grown from seed.

Planting out: Spring.

Spacing: About 18" (45 cm) apart.

GROWING

Grow these plants in **full sun** in coastal areas **or partial shade** inland. Soil should be **well drained**, and the plants may flop over in very rich soil. Catmint is good for edging borders and pathways. It blooms at the same time as roses and is often grown alongside, being one of the few plants that complement roses. Divide them in spring or fall whenever they begin to look overgrown and dense.

'Six Hills Giant'

It is no mystery where this plant gets its name—cats love it! Dried leaves stuffed into cloth toys will amuse kittens for hours.

RECOMMENDED

'Dawn to Dusk' has rose pink flowers from June to October.

'Six Hills Giant' is a large, vigorous plant, growing up to 36" (90 cm) tall and 24" (60 cm) wide. It bears large spikes of lavender blue flowers.

'Snowflake' is low-growing, compact and spreading cultivar with white flowers. It grows 12–14" (30–35 cm) tall and 18" (45 cm) wide.

N. x faassenii has many popular, usually sterile, cultivars. **'Dropmore'** has gray-green leaves and lavender blue flowers. It grows 18–24" (45–60 cm) tall and 18" (45 cm) wide.

GARDENING TIPS

Cut back catmint in June to delay flowering and make the plants more compact. Once the plants are almost finished blooming you may cut them back by one-third to one-half. This will encourage new growth and might prompt them to bloom again in late summer or fall.

Take care if you decide to grow **N. cataria** (Catnip) because cats are extremely attracted to this plant. You will be laying out a welcome mat for all the neighborhood cats to come and enjoy your garden. If you don't want cats in your garden you might want to think twice before including this species in your plans. Cats do like the other species, but not to the same extent.

PROBLEMS & PESTS

Usually pest free, except for cats or an occasional bout of leaf spot.

Long cultivated for its reputed medicinal and culinary qualities, catmint's mildly hallucinogenic qualities were also purported in the '60s.

Cinquefoil
Potentilla
Potentilla atrosanguinea var. *argyrophylla*

Flower color: Yellow, red, orange, pink or white.
Height: 12–24" (30–60 cm) tall. **Spread:** About 24" (60 cm) wide.
Blooms: Summer.

A rocky site, with poor, dry soil may seem like a hopeless spot for flowers, but Cinquefoil won't be foiled by drought, infertility or a few rocks. The small open flowers cover the plant in joyful defiance of the lousy growing conditions, and it is this great attitude that makes Cinquefoil the first choice for new rockeries or open, sunny sites.

PLANTING

Seeding: Start seeds in cold frame in spring.

Planting out: Spring or fall.

Spacing: 12–24" (30–60 cm) apart.

GROWING

Cinquefoil will grow equally well in **full sun or partial shade**—it is very drought-tolerant but should be protected from hot afternoon sun. Soil should be of **poor to average fertility and well drained**. Cinquefoil makes an attractive addition to borders, rock gardens and rock walls. Divide in spring or fall, whenever the center of the plant begins to thin or die out.

RECOMMENDED

'Firedance' grows about 15" (38 cm) tall. Its orange flowers have red centers.

'Fireflame' grows about 14" (35 cm) tall. It has bright red flowers.

'Gibson's Scarlet' has a relaxed habit and grows 12–18" (30–45 cm) tall. It has scarlet red flowers.

'Yellow Queen' grows about 12" (30 cm) tall. The flowers are yellow with red centers.

ALTERNATE SPECIES

P. nepalensis 'Miss Willmont' grows about 12" (30 cm) tall. It has strawberry-like fuzzy leaves and pink blooms that are long lasting and come true to seed.

GARDENING TIPS

Group taller types close together to support each other.

PROBLEMS & PESTS

Occasional problems with downy mildew, powdery mildew, leaf blister, rust and fungal leaf spots may occur, but good drainage usually prevents these problems.

Combine a group of Cinquefoils with a spiky yucca plant and sedums and succulents and your garden can have the 'Arizona' look despite the rain and cold of the Pacific Northwest.

Clematis

Clematis spp.

Flower color: Blue, purple, pink, red, yellow or white.
Height: 18" to 48' (45 cm to 7 m) tall. **Spread:** 24–48" (60–120 cm) or wider.
Blooms: Summer.

Imagine a Purple Smoke Tree (*Cotinus coggyria*) with burgundy leaves and the deep purple blooms of *C. viticella* 'Etoile Violette' decorating the foliage as it garlands the branches. Clematis is a plant that does a lot more than just grow up a trellis. Use it as a groundcover (it will ramble through the flower garden) or let it marry other plants together in happy union. Enjoy the gregarious nature of this plant that seems to have the urge to reach out and touch all of its neighbors.

PLANTING

Seeding: Start seeds indoors or in cold frame in late summer or fall.

Planting out: Spring or fall.

Spacing: 24–48" (60–120 cm) apart.

GROWING

Clematis prefers to be in **full sun** but will tolerate partial shade. Soil should be **fertile, humus-rich, moist and well drained**. It is an attractive plant for a trellis or the middle of the border as a specimen plant. Allow clematis to climb over your lilac shrub that has finished flowering. Divide in spring.

RECOMMENDED

There are many interesting species of clematis and dozens of hybrids.

'Jackmanii' is possibly the most well-known of all the hybrids. It grows about 10' (3 m) tall.

C. heracleifolia (Tube Clematis) grows up to 36" (90 cm) tall and 48" (120 cm) wide. The tube-shaped flowers are purple-blue. *C. h. var. davidiana* has larger, fragrant flowers.

C. recta

C. integrifolia

Clematis can be grown in a pot. Use a wire tomato cage to give it support or allow it to trail down from a hanging basket.

C. integrifolia (Solitary Clematis) grows from 18–36" (45–90 cm) tall. The plant tends to grow upwards to a point then falls to the ground and sprawls for up to 48" (120 cm). The flared bell-shaped flowers are indigo.

C. recta (Ground Clematis) is a bit more upright than the other two species. Growing up to 48" (120 cm) tall and only 24" (60 cm) wide. The fragrant white flowers are borne in dense clusters. The cultivar **'Purpurea'** has red-tinged leaves that look magnificent with the white flowers.

C. tangutica (Virgin's Bower) is a late flowering vine that can climb over 15' (5 m) tall. It has yellow pendant flowers that are followed by attractive fluffy seedheads.

C. viticella 'Etoile Violette' has purple summer blooms with a nodding habit.

C. heracleifolia

GARDENING TIPS

Shade the roots with a mulch or low shrub. Do not shade with a flat rock, which was a common practice in the past, because the rock will absorb heat and warm, instead of cool, the roots.

PROBLEMS & PESTS

Insect problems include scale insects, whiteflies and aphids. Other problems may be caused by wilt, powdery mildew, rust, fungal spots and stem cankers. To avoid wilt, keep mulch from touching the stem. Set a bottomless tuna can as a collar around newly planted clematis to help protect the fragile stem from injury. A bruised or damaged stem is a site for disease entry.

C. tangutica

The fuzzy seedheads of some species gave this plant an alternate common name— Old Man's Beard.

'Jackmanii'

Columbine

Aquilegia spp.

Flower color: Red, yellow, pink, purple, mauve, blue or white. Often the spurs will be a different, darker color than the petals. **Height:** 24–36" (60–90 cm) tall. **Spread:** 12–24" (30–60 cm) wide. **Blooms:** Spring or early summer.

Columbines are beautiful, airy plants with nodding, spurred flowers on long stems that rise up like birds in flight. Both the common and the botanical names refer to this bird-like appearance. The names are derived from the Latin words *columba* meaning 'dove' and *aquila* meaning 'eagle.' Columbines bloom in rich jewel-like colors that herald the start of summer and the passing of spring pastels.

PLANTING

Seeding: Place in cold frame or direct sow in garden in spring.

Planting out: Spring.

Spacing: About 18" (45 cm) apart.

GROWING

Columbines prefer to grow in **full sun or partial shade**. Soil should be **fertile and moist but well drained**. *A. alpina* likes to have some small gravel added to the planting hole as well, as it needs very good drainage. Columbines are one of the easiest plants to work into any garden design. They perform well in a rock garden, a formal or casual border or a naturalized woodland garden. Columbines may be divided every few years but dislike having their roots disturbed and may take a while to recover after being divided.

RECOMMENDED

McKana hybrids are available in a wide variety of colors including yellow, pink, red, purple, mauve and white.

Songbird Series are good for container gardening with larger flowers and compact foliage. **'Cardinal'** is red and **'Bluebird'** is blue.

Columbines are short-lived perennials that seed freely throughout the garden, and establish themselves in unexpected, and often charming, locations. If you wish to keep a particular form, you must preserve it carefully through frequent division or root cuttings.

A. *alpina* has blue flowers and grows 24–36" (60–90 cm) tall and 12" (30 cm) wide.

A. *canadensis* is a North American native, widespread east of the Rocky Mountains. The flowers are yellow with red spurs. It grows 24" (60 cm) tall and 12" (30 cm) wide.

A. *chrysantha* is native to southwestern North America. The flowers are yellow and it grows 36" (90 cm) tall and 24" (60 cm) wide.

A. *formosa* is a west coast native and a close relative of *A. canadensis*. The flowers are yellow with bright red spurs. It is taller than its eastern cousin, growing to be 24–36" (60–90 cm) tall and 18" (45 cm) wide.

GARDENING TIPS

Columbines self-seed, but are in no way invasive. Each year a few new plants may turn up near the parent plant. If you have a variety of columbines planted near each other you may even wind up with a new cultivar. Columbines cross-breed easily, resulting in many hybrid forms. The wide variety of flower colors is the most interesting result. The new plants may not be identical to the parents and there is some likelihood that they will revert to the original species. This won't be a problem if you are growing straight species plants. It will be if your columbines are hybrids.

PROBLEMS & PESTS

Mildew problems and rust may occur during dry summers. Fungal leaf spots are also possible. Aphids, leafminers and caterpillars are the most likely insect pests. To control leafminers, remove the leaves at first sign of damage and do a good job of fall clean-up.

In the wild A. canadensis *can be found growing in the cracks and crevices of exposed rock faces.*

Hummingbirds love columbines and are sure to visit your garden if you plant a few of these flowers.

Coral Bells

Heuchera spp.

Flower color: Red, pink or white.
Height: 12" (30 cm) tall. **Spread:** 12" (30 cm) wide.
Blooms: Spring and summer.

The tall wands with bell-shaped flowers are most often in shades of peach and coral on the common coral bells, but the deep purple leaves of *H. micrantha* var. *diversifolia* 'Purple Palace' have changed the way gardeners use this old-fashioned plant. Colorist designers, or perennial gardeners that plan specific color groupings, favor the dark foliage among the flowers as more than just an accent. It can be used next to yellow and gray foliage plants for long-lasting color and a punch of purple.

PLANTING

Seeding: The straight species may be started from seeds in cold frame in fall.

Planting out: Spring.

Spacing: 12" (30 cm) apart.

GROWING

Grow these plants in **full sun or partial shade**. Coral bells won't flower if the shade is too heavy. Soil should be **average to fertile, neutral or slightly alkaline, evenly moist and well drained**. Coral bells make good plants for edging borders and paths. They are beautiful when clustered in woodland or rock gardens. They also make good groundcovers for low-traffic or no-traffic areas. The cut flowers are interesting in arrangements. Divide in spring or fall.

RECOMMENDED

H. americana has been the parent plant of a group of new cultivars with foliage in shades of silver, purple, chocolate and dark maroon. These cultivars are beautiful and easy to grow.

H. micrantha var. *diversifolia* **'Purple Palace'** grows 18–20" (45–50 cm) tall and has bold, deep purple foliage with airy sprays of white blooms.

H. sanguinea **'Firefly'** grows up to 24" (60 cm) with densely mounded, light green foliage and red blooms on wire-like stems.

GARDENING TIPS

The spent flowers should be removed to prolong the blooming period. If the soil is acidic then horticultural lime should be applied to the soil each year. *H. sanguinea* has a strange habit of pushing itself up out of the soil. Mulch in fall if the plants begin heaving from the ground. Every two or three years they should be dug up to remove the oldest, woodiest roots and stems. Coral bells may be divided at this time, if desired, then replanted with the crown just above soil level. Cultivars may be propagated by division in spring or fall.

PROBLEMS & PESTS

Coral bells have few problems when they are healthy. They may occasionally have problems with foliar nematodes, powdery mildew, rust or leaf spots.

H. sanguinea

These delicate woodland plants will enhance your garden with their bright colors, attractive foliage and airy sprays of flowers.

'Purple Palace'

Coreopsis

Tickseed

Coreopsis spp.

Flower color: Yellow, pink or orange.
Height: 12–36" (30–90 cm) or taller. **Spread:** 12–24" (30–60 cm) wide.
Blooms: Early to late summer.

There are many types of coreopsis, and the small black seeds do look like ticks, but when viewed up close the seeds of each variety have a unique shape. This perennial is popular because it has a long blooming period, lasting most of summer, and bright, cheerful flowers that stand out above the foliage.

PLANTING

Seeding: Direct sow in garden in spring. Seeds may be sown indoors in winter, but soil must be kept fairly cool, at 55–61° F (13–16° C), in order to germinate.

Planting out: Spring.

Spacing: About 12–18" (30–45 cm) apart, according to spread.

GROWING

Grow coreopsis in **full sun**. Soil should be **average to sandy and well drained**. Coreopsis will die in moist, cool locations. Overly fertile soil causes long, floppy growth. Coreopsis is a versatile plant, useful in formal and informal borders and in a meadow planting or cottage garden. Coreopsis looks nice planted individually or in large groups. Frequent division may be required to keep plants rejuvenated.

RECOMMENDED

C. auriculata (Mouse-eared Tickseed) is low growing and well suited to rock gardens or fronts of borders. It grows 12–24" (30–60 cm) tall and will continue to creep outwards without becoming invasive.

C. verticillata

Mass plant coreopsis to fill in a dry, exposed bank where nothing else will grow, and you will enjoy the bright, sunny flowers all summer long.

C. rosea (Pink Tickseed) is an unusual species with pink flowers. It grows 24" (60 cm) tall and 12" (30 cm) wide. This species is more shade and water tolerant that the other species are. *C. rosea* is very late to break dormancy—mark it carefully to keep yourself from digging it up in the spring.

C. verticillata (Thread-leaf Coreopsis) is an attractive, mound-forming plant with long-blooming, brassy yellow flowers and finely divided foliage. It grows 24–32" (60–80 cm) tall and 18" (45 cm) wide. It is a long-lived species and will need dividing less frequently than most species. Divide it if some of the plant seems to be dying out. **'Moonbeam'** has paler yellow flowers.

C. rosea

GARDENING TIPS

Weekly deadheading will keep plants in constant summer bloom. Use scissors to snip out tall stems. Shear plants by one-half in late spring for more compact growth.

If plants blacken from frost, remove this foliage in fall to prevent slugs from using it as a nursery.

PROBLEMS & PESTS

Slugs, snails and rabbits are the most likely culprits if something has been snacking on your coreopsis. Other problems may be caused by bacterial spot, *Botrytis* flower blight, aster yellows, powdery mildew, downy mildew or fungal spots.

Coreopsis is one of the easiest perennials to start from seed. It is also one of the most gratifying because it will likely flower in the first year.

'Moonbeam'

Cornflower
Mountain Bluet
Centaurea montana

Flower color: Blue, pink or white.
Height: 18–24" (45–60 cm) tall. **Spread:** 24" (60 cm) or wider.
Blooms: Late spring to midsummer.

With a little work early in the season Cornflower will form a neat mound in a formal border, but if left to sprawl, it will also blend into an informal cottage garden. In any garden Cornflower will attract bees and butterflies with its unusual spider-like flowers. The flowers can be enjoyed fresh or in dried arrangements.

PLANTING

Seeding: Start seeds in cold frame in late summer or fall.

Planting out: Spring.

Spacing: 24" (60 cm) apart.

GROWING

Grow Cornflower in a **full sun** or **light shade** location. Any moist and well-drained soil is fine. In a rich soil, however, cornflower may develop straggly, floppy growth and may become invasive. Include these plants in the middle of mixed borders, in informal or natural gardens and in large rock gardens. Root cuttings can be taken in late winter. Division may be done about every three years in spring or late summer.

RECOMMENDED

The straight species is commonly grown because of its unusual cobalt blue flowers.

'Alba' has white flowers.

'Carnea' has pink flowers.

GARDENING TIPS

Thin the new shoots by about one third in spring to increase air circulation through the plant. Regular deadheading can extend the flowering season to last the entire summer. The invasiveness of these plants is also curbed by deadheading as it prevents the plants from self-seeding.

If the clump seems to be exending beyond the space you had intended you can use a bed edging material to surround the plant and prevent the rhizomes from extending any further.

Use a bed-edging material or a bottomless flowerpot to surround the plant and prevent it from spreading if it is becoming invasive.

PESTS & PROBLEMS

Downy or powdery mildew, rust and mold can cause occasional problems.

Corydalis

Corydalis spp.

Flower color: Yellow, blue, red or white.
Height: 12–24" (30–60 cm) tall. **Spread:** 6–18" (15–45 cm) wide.
Blooms: Various times depending on species.

The delicate foliage on this perennial is finely cut and fern-like, but, unlike ferns, the flowers are colorful and continuous. This is the longest blooming plant in my garden. The small yellow parrot-beaked flowers bloom from March until the first hard frost in October. Tidy gardeners may fret about the self-seeding habit of this enthusiastic reseeder, but I find the new plants coming up in every rock-garden crack and pathway crevice quite charming.

PLANTING

Seeding: Sow fresh seeds in fall or spring but expect erratic germination rates.

Planting out: Spring or fall.

Spacing: About 12" (30 cm) apart.

GROWING

Corydalis prefers to grow in **light shade** but will tolerate any light condition. Soil should be of **average to rich fertility and well drained**. Corydalis will tolerate moist, woodland soil. It may be grown in a light woodland garden, or rock garden or underneath shrubs in a border. Divide in spring or early summer.

RECOMMENDED

'Blue Panda' is a popular blue-flowered hybrid. It is not as invasive as *C. lutea*. The plant is compact, growing about 8" (20 cm) tall.

'China Blue' has pale blue flowers.

'Pere David' has deep blue flowers with red-tinted foliage.

C. lutea is a vigorous, even invasive, species. It has yellow flowers and may be used to fill in the space under shrubs or as a groundcover in a difficult or large area.

GARDENING TIPS

Too much sun may scorch plants in summer. Cut out yellow foliage and fresh growth will cover the plants in fall.

Corydalis loves to grow between rocks. It reseeds readily in gravel paths but seedlings are easy to pull and transplant.

PROBLEMS & PESTS

Downy mildew and rust are possible but rare disease problems. Slug and deer resistant.

The narrow, tube-like flowers of this plant may attract hummingbirds to your garden.

'Blue Panda'

C. lutea

Cranesbill Geranium
Hardy Geranium
Geranium sanguineum

Flower color: White, red, pink or purple.
Height: 8–12" (20–30 cm) tall. **Spread:** 12" (30 cm) wide.
Blooms: Summer.

These flowers blossom sporadically over most of summer in vivid colors and the foliage turns bright red in fall. Use this fantastic plant as a petticoat under tall rhododendrons or as a filler in the rock garden. Let sprawling varieties ramble through woodland gardens and use the compact types in the front of the border or rockery. Once the flowers fade the seedpods are long and pointed like a crane's bill.

PLANTING
Seeding: Certain hybrids and cultivars can be started in pots in early fall or spring.

Planting out: Spring or fall.

Spacing: 12" (30 cm) apart.

GROWING

Cranesbill Geranium prefers to grow in **full sun**, but will tolerate partial shade. Soil of **average fertility and with good drainage** is preferred, but most conditions are tolerated except waterlogged soil. These long-flowering, clump-forming plants are great in the border, filling in the spaces between shrubs and other larger plants. Divisions should be done in spring.

RECOMMENDED

'Album' has white flowers and a more open habit than other varieties.

'Alpenglow' has bright rosy red flowers and dense foliage.

'Elsbeth' has light pink flowers with dark pink veins. The foliage turns very bright red in fall.

'Shepherd's Warning' is a dwarf plant growing to 6" (15 cm) tall with rosy pink flowers.

G. s. var. striatum 'Bloody Cranesbill' grows 8–10" (20–25 cm) tall and is heat and drought-tolerant. It has pale pink blooms with blood red veins.

GARDENING TIPS

Shear back spent flowers for a second set of blooms. Ratty foliage pruned in late summer will return fresh and tidy.

The Greek word for crane is geranos lending itself to both the genus and the common name of this plant. It is the beak-like fruits of this plant that resembles a crane's bill.

PROBLEMS & PESTS

Bacterial blight, downy and powdery mildew, gray mold, leaf spot, leafminers and slugs may cause occasional problems.

G. s. var. *striatum*

Cushion Spurge
Euphorbia
Euphorbia polychroma

Flower color: Yellow
Height: 12–24" (30–60 cm) tall. **Spread:** 18–24" (45–60 cm) wide.
Blooms: Spring to summer.

This attractive mounding plant is admired for its bracts that turn bright yellow when the plant flowers. Cushion spurge provides a second show of color in the fall when the leaves change to purple, red or orange before the plant dies back for the winter.

PLANTING

Seeding: Use fresh seed for best germination rates. Start seed in cold frame in fall or spring.

Planting out: Spring or fall.

Spacing: 18" (45 cm) apart.

GROWING

Grow Cushion Spurge in **full sun** or **light shade**. Soil should be of **average fertility, moist, well drained and humus-rich**. Use Cushion Spurge in a mixed or herbaceous border, in a rock garden or in a lightly shaded woodland garden. Division can be done in late summer whenever the clump outgrows its location.

RECOMMENDED

The straight species is most commonly grown.

'Candy' has yellow bracts and flowers, but the leaves and stems are tinged with dark purple.

'Emerald Jade' is more compact and slightly smaller than the straight species. The bracts are yellow, but the flowers are green.

PROBLEMS & PESTS

Aphids, spider mites and nematodes can all cause problems for Cushion Spurge. In poorly drained soil this plant may also develop fungal rot problems.

Cushion Spurge is in the same family as E. pulcherrima, *the popular holiday Poinsettia, which has bracts in solid or speckled shades of red, cream, pink or yellow.*

GARDENING TIPS

Cushion Spurge can be propagated by stem cuttings. Dipping the cut ends in hot water before planting will stop the sticky white sap from running. You may wish to wear gloves when handling this plant because some people find the sap, which contains latex, irritates their skin.

Daylily

Hemerocallis Hybrids

Flower color: Every color except blue and true white.
Height: About 24–36" (60–90 cm) tall. **Spread:** 12–36" (30–90 cm) wide.
Blooms: Spring and summer, depending on species.

Each beautiful flower lasts only a day but daylily remains a favorite with gardeners because it covers each long stem with a multitude of gorgeous blooms. Daylilies demand attention and admiration because they won't last long enough to be taken for granted. Easygoing, adaptive and versatile, daylilies are so tough they can even be moved while in full bloom.

PLANTING

Seeding: Sow seeds in cold frame in fall or spring. Hybrids and cultivars will not come true to type from seed but can be propagated by division.

Planting out: Spring.

Spacing: 12–36" (30–90 cm) apart.

GROWING

Plant daylilies in **full sun or partial shade**. The more shade daylilies have, the fewer flowers they will produce. Though tolerant of most conditions, daylilies will produce the most flowers in a **fertile, evenly moist, well-drained** soil.

There is a place in almost every garden for this versatile plant. Daylilies may be included in a border or used for erosion control on a bank or along a ditch. Smaller types may be used in small gardens or in containers. In a natural garden daylilies may be planted in large masses or drifts. Daylilies will be more vigorous if divided every two or three years, but can be left alone for years.

Taken from the Greek words for day 'hemera' and beauty 'kallos', the genus and the common name explain that these lovely blooms last only one day.

There are gardeners who devote their lives to developing new daylily hybrids. There are already so many hybrids available that you could plant your entire garden with daylilies and no two would be alike.

RECOMMENDED

The number of forms, colors and sizes is almost infinite.

'Stella D'Oro' grows 24" (60 cm) tall and is a repeat bloomer of yellow flowers. This cultivar does well in pots.

ALTERNATE SPECIES

H. flava (formerly *H. lilio-asphodelus*) has lemon yellow flowers with a clean, sweet scent. First introduced in 1975, it is popular for its repeat blooming.

H. fulva (Tawny Daylily) grows 36–72" (90–180 cm) tall and has rusty or tawny orange blooms. This species has become naturalized across most of the country.

GARDENING TIPS

Deadheading is important to make small-flowered varieties rebloom. Purple varieties will stain fingers when deadheading.

Feed daylily twice a year in spring and summer for an abundance of blooms.

PROBLEMS & PESTS

Though daylilies are generally pest free they may be afflicted by rust, *Hemerocallis* gall midge, aphids, spider mites, thrips, slugs and snails.

Delphinium

Candle Delphinium; Candle Larkspur
Delphinium x *elatum*

Flower color: Blue, purple, pink, white or bicolors.
Height: 36–72" (90–180 cm) tall. **Spread:** 24–36" (60–90 cm) wide.
Blooms: Late spring and summer. May rebloom in fall
if deadheaded.

Beautiful blue spikes of flowers on majestically tall stems make Delphinium a vision of loveliness in photos and gardens. They can be the demanding prima donna of the perennial border, however, insisting on full sun, lots of fertilizer and careful staking. When you do coax Delphinium into bloom, be sure to save a few of the individual florets for pressing. Try drying the entire stalk by hanging it upside down in a cool dark room.

PLANTING

Seeding: May not come true to type from seed. It is best to try from purchased seed. Sow seeds directly outdoors once soil reaches 55° F (13° C) or start earlier in cold frame.

Planting out: Spring or fall with crown at soil level. It may get crown rot if planted any deeper.

Spacing: 24" (60 cm) apart.

GROWING

Grow in a **full sun** location that is well protected from strong winds. Soil should be **fertile, moist and humus-rich with excellent drainage**. Delphiniums are classic cottage garden plants. Their height and need for staking relegates them to the back of the border where they make a magnificent blue-toned backdrop for warmer foreground flowers like peonies, poppies and Black-eyed Susan. Delphiniums require division each year, in spring, to extend their lives and maintain their vigor.

These are plants so gorgeous in bloom that you can build a garden or plan a party around their flowering.

RECOMMENDED

'Belladonna' has deep, bright blue flowers and is a low, well-branched, airy cultivar with an open habit that does not require staking.

'Magic Fountain' is a series of genetic dwarfs that stand only 36" (90 cm) tall and therefore will not require staking. They come in shades of purple, white and blue.

Round Table hybrids (also known as 'Pacific Giants') are seed strains that are considered the classic Delphinium. The most popular are 'Galahad,' 'Guinivere' and 'King Arthur.'

GARDENING TIPS

The tall flower spikes have hollow centers and are easily broken if exposed to the wind. Each flower spike will need to be individually staked. Stakes should be installed as soon as the flower spike reaches 12" (30 cm) in height. You could use a wire tomato cage for a clump.

To encourage a second flush of smaller blooms remove the first flower spikes once they begin to fade and before they begin to set seed. Cut them off just above the foliage. New shoots will begin to grow and the old foliage will fade back. The old growth may then be cut right back allowing new growth. The new shoots should produce a flush of smaller flower spikes in fall.

This heavy feeder requires fertilizer twice a year in spring and summer. It loves manure mixed into soil.

PROBLEMS & PESTS

Slugs and snails are the worst problems for Delphinium. Slugs may eradicate this plant from your garden—take protective measures in early spring. Occasional problems with cyclamen mites, powdery mildew, bacterial and fungal leaf spots, gray mold, crown and root rot, white rot, rust, white smut, leaf smut and damping off are also possible.

Delphis is the Greek word for dolphin, which lends itself well to this flower: the petals of the flowers resemble the nose and fins of a dolphin.

Dwarf Plumbago

Leadwort

Ceratostigma plumbaginoides

Flower color: Deep blue.
Height: 10–18" (25–45 cm) tall. **Spread:** 12" (30 cm) or wider.
Blooms: Summer and fall bloom.

When you add this late blooming blue flower to the garden, remember where you've planted it! It will be the last to shoot up new foliage in the spring, and by that time, a new plant may have been plopped right on top of its head. My plumbago didn't seem to mind the petunia growing in its space, but stacking plants is not recommended. The small clusters of flowers are an intense blue and the leaves turn bronze and red in fall.

PLANTING

Seeding: Not recommended.

Planting out: Spring or fall.

Spacing: 12" (30 cm) apart.

GROWING

Grow Dwarf Plumbago in **full sun or partial shade**—plants will not bloom in full shade. Soil should be **average or rich and well drained**. This quick-growing plant is drought-resistant and makes an excellent and tough groundcover. Useful on exposed banks where mowing is impossible or undesirable, Dwarf Plumbago also makes a wonderful addition to a rock garden. It creeps happily between the rocks of a stone wall. Divide Dwarf Plumbago in spring.

GARDENING TIPS

Dwarf Plumbago may not die back completely in winter and it is recommended that any growth that has been killed in winter be removed in spring. Any unsightly or irregular growth may be removed in fall. Be careful not to disturb the planting site in early spring because the foliage emerges late. Cuttings may be started from new growth in early summer.

PROBLEMS & PESTS

Powdery mildew causes an occasional problem that good drainage will prevent.

Dwarf Plumbago is not actually a plumbago but is named for its resemblance to that plant.

English Daisy

Bellis perennis

Flower color: White, pink or red. All are likely to have yellow centers.
Height: 2–8" (5–20 cm) tall. **Spread:** 2–8" (5–20 cm) wide.
Blooms: Late winter to late summer.

These European natives enjoy the cool summer temperatures of the Pacific Northwest. English Daisies do have an enchanting charm when planted with spring bulbs or used at the front of a border. Perfect for giving any garden the casual sprawl of an English cottage garden, even without the bonneted fairies.

PLANTING

Seeding: Start in summer to plant in spring.

Planting out: Spring.

Spacing: 5" (13 cm) apart.

GROWING

Plant English Daisies in a **full sun or partly shady** location. Soil should be **average to rich and well drained**. Use English Daisies to edge borders or as groundcovers. They work well in rock gardens, open woodlands or in planters. Divide either in early spring or after flowering has finished.

RECOMMENDED

'Dresden China' has light pink, double flowers.

'Monstrosa' is 8" (20 cm) tall with red, double flowers that are 3" (8 cm) wide.

'Shrewly Gold' has gold-toned leaves and single flowers.

'White Pearl' has white, double flowers.

GARDENING TIPS

English Daisies have the habit of self-seeding and are prone to show up where you least expect them, including in your lawn. Deadheading to control spread is possible but decidedly time consuming because English Daisies are low-growing. Being a personal fan of flowering plants in lawns, I simply let these small and attractive flowers be. However, if immaculate lawns are required, place this species in beds that are well away from lawns and consider taking the time to deadhead.

PROBLEMS & PESTS

English Daisies are occasionally afflicted with fungal leaf spots. Aphids may also be a problem.

Children in England are often told that garden fairies use these small pink daisies as caps on rainy spring days.

English Daisies invading a lawn.

Evening Primrose
Sundrops
Oenothera spp.

Flower color: Yellow, pink or white.
Height: 12–36" (30–90 cm). **Spread:** 8–12" (20–30 cm) or wider.
Blooms: Spring and summer.

The large papery blossoms on this plant look like they should be blooming on a cactus, and so it is no surprise that the evening primrose is a sun-loving, drought-tolerant plant. Originally, all varieties had flowers that were held tightly closed until the afternoon sun had set, but now there are many varieties that bloom during the day. This is a plant that thrives in poor, rocky soil and blooms better because of it.

PLANTING

Seeding: Sow seeds in cold frame in early spring.

Planting out: Spring.

Spacing: About 12" (30 cm) apart.

GROWING

Grow these plants in a **full sun** location. Soil should be **poor to average, very well drained** and even a bit stony. Use *Oenothera* in the front or middle of a border. The lowest growing species may be used to edge a border or sunny path or to brighten a gravelly bank or rock garden. Divide in spring, when needed.

RECOMMENDED

O. fruticosa (Sundrops) has deep yellow flowers and leaves that turn red after a light frost. It grows 12–36" (30–90 cm) tall and 12" (30 cm) wide. The flowers are borne on spikes in groups of three to ten. The flowers open during the day. **'Summer Solstice'** (or **'Sonnenwende'**) is a more compact plant with larger, longer blooming flowers. The leaves turn red in summer and burgundy in fall.

O. macrocarpa (Ozark Sundrops) is a 10" (25 cm) tall, lower growing species with large yellow flowers up to 5" (13 cm) across.

O. speciosa (Showy Evening Primrose) has white or pink flowers that open during the day. It grows 12" (30 cm) tall and wide.

GARDENING TIPS

These plants may become invasive if the soil is too fertile. Mix gravel into the soil where they will be planted to be sure they will have adequate drainage, particularly in winter.

PROBLEMS & PESTS

Downy and powdery mildew, leaf gall, rust and leaf spots are possible problems. Roots may rot if the wet winter soil doesn't have adequate drainage.

O. speciosa

Another common name for this plant is evening star, because at night the petals emit phosphorescent light.

O. fruticosa

False Solomon's Seal
Treakleberry
Smilacina racemosa

Flower color: White.
Height: About 36" (90 cm) tall. **Spread:** 24" (60 cm) wide.
Blooms: Mid- or late spring.

Called False Solomon's Seal because it resembles true Solomon's Seal, this tall shade-loving perennial will remind you of a ballerina as its gracefully arching stems bend and sway in the breeze. The white blooms are attractive in spring and the berries are a lovely red with purple spots in fall.

PLANTING

Seeding: Start seeds in cold frame in fall, but they may take years to flower.

Planting out: Spring or fall.

Spacing: 24" (60 cm) apart.

GROWING

False Solomon's Seal prefers **light or full shade.** Soil should be of **average fertility, humus-rich, acidic, moist and well drained.** This plant is ideal for open woodland plantings and in natural gardens. Divide in spring.

GARDENING TIPS

Add peatmoss to soil at planting time to give this plant the highly organic and slightly acidic soil it loves.

PROBLEMS & PESTS

Occasional problems with rust and leaf spots are possible. Slug and deer resistant.

The berries of this plant ripen in an unusual way. The unripe green berries develop little red spots that eventually cover the entire fruit.

Fleabane

Erigeron spp.

Flower color: White, pink, purple, orange or yellow.
Height: 4–24" (10–60 cm) tall. **Spread:** 9–24" (23–60 cm) wide.
Blooms: Summer.

Carefree wildflowers that will bloom even if your soil is a bit on the sandy side, fleabane got its name because it was once used to keep fleas from the cabin. The cure is not something you would want to try today however, as it involved burning the plant inside the room so that smoke filled the space thickly, but it was still possible to breathe. Enjoy fleabane today with other late summer flowers in the rock garden or border.

PLANTING

Seeding: Start in cold frame in mid- or late spring.

Planting out: Fall.

Spacing: 9–24" (23–60 cm) apart.

GROWING

Fleabane will grow in **full sun or light shade**. Soil should be **average to rich and well drained**, even sandy. The plants will produce more flowers if the soil does not become excessively dry. Occasional supplemental watering during dry spells may be required. Fleabane is an excellent choice for the middle or foreground of a border. Low, mat-forming species are attractive when used in a rock garden or on a rock wall. Divide fleabane every two or three years in spring or fall. Discard the old woody growth when dividing.

'Canary Bird'

E. glaucus

Fleabane can be used in fresh flower arrangements. Wait until the flowers are completely open before cutting them to make your arrangement last longer.

RECOMMENDED

E. aureus **'Canary Bird'** is a low mat-forming species, native to the Rocky Mountains. It rarely grows any larger than 4" (10 cm) tall and 9" (23 cm) wide. The flowers are a bright canary yellow.

E. glaucus (Beach Aster) is a sprawling plant, native to the Pacific coast. It grows about 12" (30 cm) tall and 18" (45 cm) wide. Good cultivars are available in pink or white, both with yellow centers. It is an excellent addition to a coastal garden, being quite tolerant of salt spray.

E. speciosus (Oregon Fleabane) is another native to the Pacific coast. Many cultivars and hybrids have been developed from this species. They range in size from 12–24" (30–60 cm) in height and width. A few cultivars are **'Charity'** with light pink flowers, **'Darkest of All'** with semi-double, dark purple flowers, **'Dignity'** (a smaller cultivar) with deep pink flowers and **'White Quakeress'** with very pale, almost white flowers. All flower centers are yellow.

'Canary Bird'

GARDENING TIPS

Deadheading these plants will encourage them to continue flowering. These plants are notoriously long-lived and regular division will keep them vigorous. The taller species require support to prevent them from flopping over. Twiggy branches pushed into the soil around the plants while they are small will provide support as they grow and will be hidden once the plant is mature.

PROBLEMS & PESTS

Downy and powdery mildew, rust, smut and leaf spots occasionally occur. Prevent mildew problems by thinning out the new growth in early spring and pruning the stems back in late spring. These measures will encourage compact growth and improve air circulation.

E. speciosus

Foamflower
Sugar-Scoop
Tiarella spp.

Flower color: White or pink.
Height: 4–24" (10–60 cm) tall. **Spread:** 12–18" (30–45 cm) wide.
Blooms: Late spring.

Foamy pink or white flowers give this shade-loving woodland plant its name, and the heart-shaped leaves add to the storybook charm of this easy to grow groundcover. Don't be afraid of foamflower taking over the garden because it spreads rather politely by underground stolons and is easily pulled up if it wanders too enthusiastically. It is an attractive plant that looks so natural under trees that you'll think it was planted by Mother Nature herself.

PLANTING

Seeding: Start seeds in cold frame in spring.

Planting out: Spring.

Spacing: 12–18" (30–45 cm) apart.

GROWING

Foamflower prefers **partial or full shade** avoiding afternoon sun. Soil should be **humus-rich, moist and slightly acidic.** Foamflower is an excellent groundcover for shaded and woodland gardens. Divide in spring. This plant spreads by underground stolons, which are easily pulled up to stop the plant from spreading.

RECOMMENDED

'Maple Leaf' forms a mound of red-flushed leaves and has white, pink-tinged flowers. It grows about 12" (30 cm) tall and wide.

T. cordifolia grows 4–12" (10–30 cm) tall and 12" (30 cm) wide. The flowers are white, often tinged with pink. Native to eastern North America, it will spread freely but not invasively.

T. laciniata grows 12" (30 cm) tall and 18" (45 cm) wide. The white flowers are borne in summer. It is a Northwest native.

T. trifoliata grows 20" (50 cm) tall and 12" (30 cm) wide. The white flowers are pendant and borne in late spring through to midsummer. It is another Northwest native. A pink-flowered cultivar called **'Incardine'** is available.

PROBLEMS & PESTS

May have problems with slugs and rust.

GARDENING TIPS

Deadheading will encourage reblooming. If foliage fades or rusts in summer, cut it partway to the ground and fresh new growth will appear.

The starry flowers cluster along the long stems looking like festive sparklers.

Forget-Me-Not

Myosotis spp.

Flower color: Blue, also pink, yellow or white. All have white or yellow eyes.
Height: 6–12" (15–30 cm) tall. **Spread:** 6–12" (15–30 cm) wide.
Blooms: Spring through late summer.

The tiny baby blue and pastel pink blooms appear just in time to celebrate all the new spring babies. My cat once had a litter of kittens that loved pouncing in my bed of forget-me-not—the sweet sight won't be forgotten.

It is a delightful addition to woodland or wet areas and a great companion to wildflower and native plant gardens.

PLANTING

Seeding: Start seeds in cold frame in spring or let self-seed.

Planting out: Spring or fall.

Spacing: 6–12" (15–30 cm) apart.

GROWING

Forget-me-not prefers to be located in **light or partial shade** but will tolerate full sun if it has plenty of water. Soil should be **average to poor in fertility, moist and well drained**. Water Forget-me-not may be used along the edges of a stream or pond where it will self-seed and naturalize. Woodland Forget-me-not may be used in a shady border, as a low-growing woodland plant or in a rock garden. It will also self-seed. Divide in spring or fall if desired.

RECOMMENDED

M. scorpioides (Water Forget-me-not) bears clusters of bright blue flowers with white, yellow or pink eyes. It may grow up to 12" (30 cm) tall and wide. It may even be grown directly in shallow water.

M. sylvatica (Woodland Forget-me-not) is usually grown as a biennial. It bears clusters of blue or white flowers, with cultivars also available in shades of pink. It may grow up to 12" (30 cm) tall, but only 6" (15 cm) wide. This plant also looks good when grown in combination with other wildflowers.

GARDENING TIPS

These are short-lived perennials that will self-seed. Allow seeds to set to provide you with future generations of plants. After they bloom, plants may blacken and dry up. Pull these up and lay them wherever you want a new colony to grow next.

PROBLEMS & PESTS

Snails, slugs, powdery and downy mildew, rust and gray mold may all cause problems.

The name comes from the way this biennial lives a short life after blooming but then reappears as new seedlings all over the garden.

Foxglove
Digitalis purpurea

Flower color: Pink, purple, yellow, maroon, red or white.
Height: 36–72" (90–180 cm) tall.
Spread: 24" (60 cm) wide. **Blooms:** Summer.

Imagine the small red fox, slipping through country gardens, wearing nothing but foxglove florets as slippers. The spotted flowers are just the right size to slip over the fingers of a child's hand. They also attract hummingbirds, which poke their tiny bodies halfway up into the bell-shaped blooms. These short-lived perennials have the pleasant habit of self-seeding, providing a constant supply of new plants to replace the ones that die out.

The extremely poisonous nature of foxglove has been known to cause rashes, headaches and nausea simply from touching the plant.

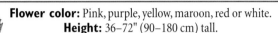

PLANTING

Seeding: Direct sow in garden or start in pots in cold frame. Flowers unlikely in first year.

Planting out: Spring.

Spacing: 18" (45 cm) apart.

GROWING

Foxgloves prefer a **partially shady** location and a soil that is **fertile, humus-rich and acidic**. They will tolerate almost any soil condition other than overly wet or overly dry soil. Foxgloves are a must-have for the cottage garden or for those people interested in heritage plants. They make an excellent vertical accent along the back of a border. They are also an appropriate addition to a woodland garden because they naturally occur in woodland areas.

The heart medication digitalis is made from extracts of foxglove. For over 200 years Digitalis *has been used for treating heart failure.*

Division is unnecessary because these plants will not live long enough to be divided. They continue to occupy your garden by virtue of their ability to self-seed. If too many plants are growing, you may wish to thin them out or transplant some to another location—perhaps into a friend's garden.

RECOMMENDED

'Alba' has white flowers.

'Excelsior' comes in a wide range of colors with densely held flowers.

'Foxy' is a dwarf plant growing to a maximum height of 36" (90 cm) in a wide range of colors.

Hybrid varieties may become less vigorous over time. Sowing newly purchased seeds into the bed each year will help prevent this problem.

ALTERNATE SPECIES

D. x mertonensis (Strawberry Foxglove) is a true perennial, unlike most fox-glove cultivars. It has rose pink blooms and grows 36–48" (90–120 cm) tall.

GARDENING TIPS

Some staking may be required, if the plants are in a windy location. This is another plant that you may wish to deadhead once it is finished flowering. The drawback to this is that in a few years' time you may no longer have Foxglove in your garden. Leave some of the flower spikes in place to spread seeds to allow for new plants. Remove the tallest spike and side shoots will bloom.

PROBLEMS & PESTS

Anthracnose and fungal leaf spot may be problems. Slug and deer resistant.

Gayfeather

Blazing Star

Liatris spicata

Flower color: Red-purple to purple, sometimes white.
Height: About 20–60" (50–150 cm) tall in bloom. **Spread:** 18" (45 cm) wide.
Blooms: Summer.

Grassy clumps of foliage make this spiky plant look great next to smooth rocks and boulders. In summer, when the unusual spikes of pink or purple flowers rise from the center of the plant, it reminds me of fireworks lighting up the summer sky. A great cut flower when used fresh for line designs in modern-styled arrangements, the blooms are also easy to dry for enjoyment all winter long.

PLANTING

Seeding: Start seeds in cold frame or direct sow in garden in fall. May not flower for two to four years.

Planting out: Spring.

Spacing: 24" (60 cm) apart.

GROWING

Grow these plants in **full sun**. Soil should be **average or sandy, with plenty of humus** mixed in. They prefer to be well watered while they are growing. Use Gayfeather in a border, meadow planting or in a bog garden. Mulch well during summer to prevent moisture loss when the plant would prefer to be moist. Divide every three or four years in fall if the clump is becoming crowded.

RECOMMENDED

'**Floristan White**' is 36" (90 cm) tall with white flowerheads.

'**Kobold**' is 20–30" (50–75 cm) tall with deep purple flowerheads.

'**Snow Queen**' has white flowerheads.

GARDENING TIPS

Trim off the spent flower spikes to promote a longer blooming period and to keep the plants looking tidy.

Gayfeather doesn't like to sit in water during cool weather and may develop root rot during winter. Plant in a location that has good drainage. Gayfeather will do fine in pots.

PROBLEMS & PESTS

Slugs and snails as well as stem rot, rust and leaf spots are the most common problems.

Gayfeather is an interesting and beautiful plant to add to your collection of moisture-loving pondside plants. The fluffy flower spikes are quite unique, particularly when reflected in the water, giving the impression of stars shooting in all directions.

Gentian

Gentiana spp.

Flower color: Blue most common, also white, yellow, pink or purple.
Height: 6–36" (15–90 cm) tall. **Spread:** 8–18" (20–45 cm) wide.
Blooms: Late summer and fall.

Gentian blue is the most intense, bright navy blue you will find in the natural world, and the Pacific Northwest is one of the few places, other than mountainsides, where gentians can be grown. Keeping these temperamental plants happy is not an easy task, but the challenge of capturing a piece of the sky to bloom from the ground is reward enough for dedicated rock gardeners.

PLANTING

Seeding: Start ripe seeds in cold frame.

Planting out: Spring.

Spacing: 6–12" (15–30 cm) apart.

GROWING

Grow these plants in **full sun or partial shade**. Soil should be **light, humus-rich and evenly moist with good drainage**. Gentian prefers soil to be neutral or acidic. It is a beautiful, late-flowering addition to borders. Low-growing species are good additions for the rock garden. Cultivars can be propagated from root offsets or plant division. Divide in spring.

RECOMMENDED

G. asclepiadea (Willow Gentian) grows about 24–36" (60–90 cm) tall and 18–24" (45–60 cm) wide with dark blue blooms.

G. sceptrum is a native of the Pacific coast. It grows 18–36" (45–90 cm) tall and 8–12" (20–30 cm) wide.

G. septemfida (Crested Gentian) grows 6–12" (15–30 cm) tall and 12" (30 cm) wide and is the least demanding of the gentians to grow.

PROBLEMS & PESTS

Rust and fungal leaf spots are the most common problems. Slugs and snails may also cause occasional problems.

G. asclepiadea

GARDENING TIPS

Mulch and add peat moss to the planting hole to keep roots cool and to make soil even more acidic. This plant does not like to be transplanted and is slow to establish itself.

The roots of some gentian species have medicinal benefits, generally as digestive bitters. The taste of gentian is so bitter that the flavor is still detectable when diluted to one part in 12,000.

G. asclepiadea

Geum
Avens
Geum spp.

Flower color: Orange, red, yellow, cream or pink.
Height: 6–24" (15–60 cm) wide. **Spread:** 12–24" (30–60 cm) wide.
Blooms: Spring and early summer.

Flowers bloom in bright sunset shades on these mid-height plants, which makes them great standouts in the middle of a perennial border. A native of the cool mountains of South America, geum wants a cool root run, which means a layer of mulch, and good air circulation will keep them pushing up the budded stems.

There are species of geum found in mountainous regions all over the world.

PLANTING

Seeding: Direct sow freshly ripened seeds in fall.

Planting out: Spring or fall.

Spacing: 12–24" (30–60 cm) apart.

GROWING

Plant in a location that receives **full sun** but doesn't get too hot. Soil should be **fertile, evenly moist and well drained**. Geum does not like water-logged soil. Geum makes a bright-flowered addition to the border. It looks particularly attractive when combined with plants that have dark blue or purple flowers. Divide geum in spring or fall. This plant can be short-lived; divide frequently to increase longevity.

RECOMMENDED

G. coccineum (Scarlet Avens) bears bright scarlet red flowers from late spring to late summer. It grows 12–20" (30–50 cm) tall and 12" (30 cm) wide. **'Borisii'** has fine foliage and hot orange flowers.

G. quellyon also known as *G. chiloense* (Chilean Avens) bears scarlet flowers all summer. It grows 16–24" (40–60 cm) tall and 24" (60 cm) wide. **'Lady Stratheden'** has golden yellow flowers. **'Mrs. Bradshaw'** has orange-red flowers.

G. coccineum

GARDENING TIPS

Cut off spent flowers to keep more coming. You will need to divide every year in spring or late summer. Place gravel in the planting hole.

PROBLEMS & PESTS

Geum may have problems with downy mildew, powdery mildew, fungal leaf spots and leaf smut. Caterpillars occasionally cause trouble.

G. quellyon

Goat's Beard

Aruncus dioicus

Flower color: Cream.
Height: About 72" (180 cm) tall. **Spread:** About 48" (120 cm) wide.
Blooms: Summer.

Goat's Beard will remind you of the white, fluffy beards of the mountain goats in the alpine wilderness. Luckily you won't need the alpine climate and can grow this graceful but tough plant with fluffy white flowers in your garden. Enjoy the arching branches and use this large perennial shrub as a background or transition plant—and think of mountain goats every time it blooms.

PLANTING

Seeding: Direct sow once soil is warm, or in cold frame in spring or fall. Seeds should sprout within two or three weeks.

Planting out: Spring or fall.

Spacing: 48" (120 cm) apart.

GROWING

Goat's Beard prefers to grow in **partial to full shade**. Planted in a deep shade location will cause the plant to have fewer blooms. It will also tolerate

The attractive foliage and flowers and the large size of Goat's Beard make it a popular plant for shade gardens.

Goat's Beard looks a bit like Astilbe and is often grouped with this smaller plant to create an interesting contrast in size.

full sun as long as the soil is kept evenly moist. Soil should be **rich and moist, with plenty of humus** mixed in. This plant looks very natural growing along the edge of a woodland or in an open forest. It may be used at the back of a border or alongside a stream or pond. Division should be done in spring or fall. Goat's Beard may be quite difficult to divide because it develops a thick root mass. Use a sharp knife to cut the root mass into pieces.

RECOMMENDED

'Kneiffii' grows 36–48" (90–120 cm) tall and 18" (45 cm) wide. This cultivar is very dainty with finely divided leaves, arching stems and nodding plumes of flowers.

'Zweiweltkind' (meaning 'child of two worlds') is a compact plant with drooping, pendulous, white flowers.

A. d.* var. *astilbioides is a dwarf variety. It grows to only 24" (60 cm) tall.

GARDENING TIPS

Goat's Beard has both female and male plants, both bearing flowers. The male flowers are full and fuzzy while the female flowers are pendulous.

Goat's Beard tends to self-seed, but it is recommended to remove the spent flowers to maintain the attractive appearance of the plant and encourage a longer blooming period. If you want to start some new plants from seed then allow the seedheads to ripen before removing them. Keep in mind that you will need to have both male and female plants in order to produce seeds that will sprout. Don't save male flowerheads; they will not produce seeds.

PROBLEMS & PESTS

Occasional problems with fly larvae or tarnished plant bugs are possible.

Golden Marguerite

Anthemis tinctoria

Flower color: Yellow, cream white or white.
Height: 24–36" (60–90 cm) tall. **Spread:** 24–36" (60–90 cm) wide.
Blooms: Late spring to late summer.

There is no flower as simple and pure as the daisy. Growing lots of Golden Marguerite will give any garden a casual cottage look. Teach a child to make a daisy chain or pluck the petals to discover true love. This is a fun flower to tuck into a vase, as daisies look comfortable soaking up water in anything from tin cans to old tea pots.

PLANTING

Seeding: Plant seeds outdoors in spring.

Planting out: Spring.

Spacing: 18–24" (45–60 cm) apart.

GROWING

Marguerites prefer a **full sun** location. Soil may be **average to poor** and should be **well drained**. This plant is drought-tolerant. Marguerites form attractive clumps that blend wonderfully into a cottage-style garden. Their drought tolerance makes them ideal for use in rock gardens and on exposed slopes. Marguerite clumps tend to die out in the middle and should therefore be divided every two or three years in spring or fall in order to keep them looking their best.

RECOMMENDED

'Beauty of Grallach' is a deep orange-gold. This cultivar has deep flower colors.

'E.C. Buxton' has cream flowers with yellow centers.

'Grallach Gold' has bright yellow flowers.

'Moonlight' has the largest flowers among the varieties. They are a light yellow color.

GARDENING TIPS

These plants will tend to flower in waves. Cutting off the dead flowers will encourage continual flowering all summer. If the plant begins to look thin and spread out you may cut it back hard to promote new growth and flowers. To avoid the need for staking, group several plants together so they can support each other. If stakes are used then small, twiggy branches, installed while the plant is small, will support the plant and be hidden once the plant gets taller.

PROBLEMS & PESTS

Marguerites may get fungal problems, such as powdery or downy mildew, but cases are rare. They are most likely to have trouble with aphids or slugs.

When the botanical name of a plant includes the word tinctoria *it refers to the use of the plant for dyeing. A yellow dye can be extracted from the flowers of* A. tinctoria.

Hens and Chicks
Houseleek
Sempervivum tectorum

Flower color: Red, yellow, white or purple.
Height: 6" (15 cm) tall. **Spread:** 20" (50 cm) wide. May take a while to spread.
Blooms: Summer.

In my own garden, Hens and Chicks nest contentedly in a pair of old leather boots that once belonged to my gardening grandfather. I filled the boots with soil, cut holes in the shoe leather at the toes and planted the succulents many years ago. They have never been replanted, but new generations of chicks continue to replace the oldest hens that poke from the tops and sides of the well-worn boots. This is a plant that will grow just about anywhere it can get some sun and well-drained soil.

The juice from the leaves has astringent qualities that can be applied to burns, insect bites and other skin problems.

PLANTING

Seeding: Start seeds in cold frame in spring.

Planting out: Spring.

Spacing: 10" (25 cm) apart.

GROWING

Grow these plants in **full sun or partial shade**. Soil should be **poor to average and very well drained**. Add fine gravel or grit to the soil to provide adequate drainage. These plants make excellent additions to rock gardens and rock walls, where they will even grow on the rocks. Divide by removing new rosettes and rooting them. The whole plant may be left in place and the old rosettes removed periodically to provide space for the new rosettes.

RECOMMENDED

'Artropurpureum' has dark reddish-purple leaves.

'Limelight' has yellowish-green, pink-tipped leaves.

'Pacific Hawk' has dark red leaves with white, hairy edges.

'Silverine' has silvery blue-gray leaves edged and tipped with pink.

ALTERNATE SPECIES

S. arachnoideum has hairy threads that connect the tips of all the leaves, looking like cobwebs.

GARDENING TIPS

These plants can grow on almost any surface. In the past they were grown on tile roofs. It was believed they would protect the house from lightning.

The biggest challenge faced when growing these plants is preventing them from getting too wet in winter. If you don't have a well-drained location in the garden, growing them under an overhang of the house or in clay containers that can be moved onto a sheltered balcony in winter will help. The hairy-leafed species are more likely to be killed by too much moisture.

PROBLEMS & PESTS

Usually pest free. May have rust problems and rot in wet soil.

Himalayan Poppy

Meconopsis betonicifolia

Flower color: Sky blue, purple or pink.
Height: About 48" (120 cm) tall. **Spread:** 18" (45 cm) wide.
Blooms: Late spring and early summer.

A sea of these blue poppies is such a soothing sight that a Northwest show garden and former estate uses the Himalayan Poppy as the signature flower for its brochures and promotions. This poppy is a bit difficult to grow, with specific growing requirements, but when one fine, fat bud finally opens successfully, unfurling tissue-paper-thin petals of sky blue, that day in May will be one to celebrate in your garden.

PLANTING

Seeding: Start seeds in cold frame in fall or direct sow uncovered in garden in late summer or following spring.

Planting out: Fall.

Spacing: About 12" (30 cm) apart.

GROWING

Himalayan Poppies grow well in a **sheltered, partially shaded** site. Soil should be **fertile, humusrich, moist, well drained and slightly acidic**. Blue poppies look wonderful when mass-planted. They look interesting when planted under a stand of tall rhododendrons or in a woodland garden. These are short-lived perennials that won't need dividing.

ALTERNATE SPECIES

M. cambrica (Welsh Poppy) is a western European species with bright yellow flowers. It grows 12–18" (30–40 cm) tall and spreads 10–12" (25–30 cm) wide. It enjoys the same growing conditions as the Himalayan Poppy but is more tolerant of wet winter conditions.

GARDENING TIPS

These plants self-seed quite easily, which will provide you with more plants as the old ones die out. They may take two to four years to bloom from seed and if they have been allowed to dry out too often they may die after they are finished flowering. New seedlings growing in will keep you supplied with new plants.

To help your plants live a longer life it is a good idea to remove the flower buds until the plant has developed at least three or four crowns at the base.

M. cambrica

M. cambrica

PROBLEMS & PESTS

Slugs and snails as well as downy mildew and damping off are possible problems.

Some species of Meconopsis *are referred to as monocarpic perennials. This means that they may live for several years, but that they only flower once before dying.*

Hollyhock

Alcea rosea

Flower color: Yellow, white, apricot, pink, red, purple or reddish black.
Height: 60–96" (150–245 cm) tall. **Spread:** 24" (60 cm) wide.
Blooms: Summer and sometimes into early fall.

Nothing says 'storybook charm' like Hollyhocks against a picket fence or stone wall, and using them in the back of the border up against a wind break is also the most practical way to grow these 'grandma flowers.' The best and healthiest Hollyhocks I ever saw were growing wild in a back alley. Seedlings that escape from a garden and make it on their own are often the most disease-resistant plants.

Hollyhock was originally grown as a food plant. The leaves were added to salads.

PLANTING

Seeding: Start seeds in fall—indoors for flowers in first year or outdoors for flowers in second year. Harvest seeds and scatter on ground in fall.

Planting out: Spring or fall.

Spacing: 24–36" (60–90 cm) apart.

GROWING

Hollyhocks prefer to grow in **full sun** but will tolerate partial shade. Soil should be **average to rich and well drained**. Because they are so tall, Hollyhocks look best at the back of the border or in the center of an island bed. If placed up against a fence they will get some support from the fence. In any windy location they will need to be staked. New daughter plants that develop around the base of the mature plant may be divided off in order to propagate fancy or desirable specimens. They don't always come true to type from seed, so this may be the only way to get more of a plant that is particularly attractive.

The powdered roots of plants in the mallow family, to which Hollyhock belongs, were once used to make a soft lozenge for sore throats. Though popular around the campfire, marsh-mallows no longer contain the throat soothing properties they originally did.

RECOMMENDED

'Chater's Double' bears double flowers in all shades.

'Majorette' is a smaller cultivar, attaining a height of only 36" (90 cm) and width of only 12" (30 cm). The flowers are fringed and semi-double in shades of yellow, deep red or apricot pink.

'Nigra' has single flowers in dark reddish black with yellow throats.

GARDENING TIPS

Old-fashioned Hollyhocks typically have single flowers and grow much taller than newer hybrids but are more disease resistant. Growing them as biennials and removing them after they flower is a good way to keep rust at bay.

Collect seeds and start new plants in different parts of the garden to have a display every year.

Hollyhocks may be grown shorter and bushier with smaller flowers if the main stem is pinched out early in the season. These shorter flower stems are less likely to be broken by the wind and therefore can be left unstaked.

'Nigra'

PROBLEMS & PESTS

Hollyhock rust is the biggest problem. Hollyhocks are also susceptible to bacterial and fungal leaf spots. Slugs and cutworms occasionally attack young growth. Sometimes mallow flea beetles, aphids or Japanese beetles may cause trouble.

'Chater's Double'

Hosta

Plantain Lily

Hosta spp.

Flower color: White or purple.
Height: 6–36" (15–90 cm) tall. **Spread:** 12–36" (30–90 cm) wide.
Blooms: Summer and fall.

Large, broad and colorful leaves makes hosta an excellent filler, background or specimen plant, and the variety of color and shape available will dazzle in your garden.

PLANTING

Seeding: Start seeds in cold frame in spring.

Planting out: Spring.

Spacing: 12–36" (30–90 cm) apart.

GROWING

Hostas prefer to be planted in **filtered sunlight or partial shade**. If they will be getting full sun during some part of the day it should be in the morning. Soil should be **fertile, moist and well drained**. Hostas are fairly drought-tolerant, especially if given a mulch to help retain moisture.

Hostas make wonderful woodland plants, looking very attractive when combined with ferns and other fine-textured plants. Hostas are also good to use in a mixed border, particularly when used to hide the lower stems and branches of shrubs that tend to lose their lower leaves. Division is not necessary but may be done, in spring or early summer, every few years if more plants are desired.

Hostas are considered by some to be the ultimate in shade plants. They are available in a wide variety of leaf shapes and colors.

RECOMMENDED

There are hundreds of varieties of hosta. The following are a couple of dependable favorites.

'Royal Standard' is 36" (90 cm) tall with fragrant flower spikes. It is tough and durable.

H. fortuneii **'Francee'** is 15–18" (38–45 cm) tall with dark green, white-margined leaves.

GARDENING TIPS

Hosta flowers are attractive and fragrant. However, the flower color often clashes with the leaves, which are the main decorative feature of the plant. If you don't like the look of the flowers feel free to remove the them before they open—this will not harm the plant.

PROBLEMS & PESTS

Slugs and snails are the worst problems for hosta. Some slug-resistant varieties are available. Keep an eye open for intruders. Saucers of beer and rings of diatomaceous earth around the plants may help keep slugs and snails away. Watch for green cabbage worms in late summer. You can distinguish worm damage from that of slugs by the dark droppings the worms leave on the foliage.

Some hosta gardeners feel that the leaves do not develop their best coloring and shape unless the plant is left undivided for many years. These gardeners detach small segments of the plant clump to propagate new plants without digging up the entire plant or disturbing the roots.

Iris
Iris spp.

Flower color: Shades of pink, red, purple, blue, white, brown or yellow.
Height: Varies from under 6" to over 48" (15–120 cm) tall.
Spread: Varies from 2–48" (5–120 cm) wide. **Blooms:** Spring and summer.

Spiky, grassy leaves emerge like swords from the spring ground, all set to protect the complex, jointed flowers. A flower that blooms in true blue as well as rich jewel tones and soft pastels, iris comes from such a varied family of plants that any garden can enjoy its charms. An old-fashioned flower much loved by Victorians for the ruffled petals and ornate trims, iris may bring back childhood memories of Grandma's spring garden.

PLANTING

Seeding: Start seeds in cold frame in fall.

Planting out: Late summer or early fall.

Spacing: 2–48" (5–120 cm) apart.

GROWING

Irises prefer to grow in **full sun** but will tolerate very light or dappled shade. Soil should be **average or fertile and well drained**. Irises are popular border plants, but some species are also useful along a stream or pond and others are attractive additions to rock gardens. Divide in late summer or early fall.

RECOMMENDED

I. douglasiana is native to the Pacific Northwest and is well suited to the conditions there. Many colors are available. The plant size ranges from 6–28" (15–70 cm) tall. They flower in late spring and early summer.

I. germanica (Bearded Iris) produces flowers in all colors. This iris has been used as the parent plant for many desirable cultivars. Cultivars may vary in height and width from 6–48" (15–120 cm). Flowering periods range from midspring to midsummer and some cultivars flower again in fall.

I. innominata is native to the Pacific Northwest. It grows 6–10" (15–25 cm) tall. Flower colors range through shades of yellow and purple, sometimes with veining darker than the petal color.

I. siberica (Siberian Iris) is more resistant to Iris borers than other species. It grows 24–48" (60–120 cm) tall and 36" (90 cm) wide. It flowers in early summer and there are cultivars available in many shades—mostly purple, blue and white.

I. unguicularis (Algerian Iris) provides color to winter gardens with flowers in shades of purple. The plant grows about 12" (30 cm) tall. It flowers in late winter, early spring and sometimes in late fall.

I. germanica

The iris is depicted on the wall of an Egyptian temple dating from 1500 BC, making it one of the oldest cultivated plants.

Iris, meaning 'rainbow' in Greek, was named after the Greek goddess of the rainbow. The name relates to this flower's variable colors.

Pacific Coast hybrids are a group of popular and attractive cultivars developed mostly through crosses between *I. douglasiana* and *I. innominata*. Many colors and sizes are available.

GARDENING TIPS

It is a good idea to wash your hands after handling these plants because they can cause severe internal irritation if ingested. Make sure they are not planted close to places where children play.

Pacific Coast hybrid

PROBLEMS & PESTS

There are several problems that are quite common to irises. Close observation will prevent these problems from becoming severe. Iris borers are a lethal problem. They burrow their way down the leaf until they reach the root where they continue eating until there is no root left at all. The tunnels they make in the leaves are easy to spot, and if infected leaves are removed and destroyed, or the borers squished within the leaf, they will never reach the roots.

Leaf spot is another problem that can be controlled by removing and destroying infected leaves. Be sure to give the plants the correct growing conditions. Too much moisture for some species will allow rot diseases to settle in and kill the plants. Slugs, snails and aphids may also cause some trouble.

Powdered iris root, called orris, smells like violets when crushed and was added to perfumes and potpourris as a fixative.

I. siberica

Jacob's Ladder

Polemonium spp.

Flower color: Blue, purple, pink, yellow or white.
Height: 4–36" (10–90 cm) tall. **Spread:** 8–12" (20–30 cm) wide.
Blooms: Spring and summer.

Jacob is a character from the Bible who dreamt of a ladder leading up to heaven. The precise arrangement of the horizontal leaves presents an image of nature's ladder leading to a sky blue heaven of sweetly scented blooms. The fragrance of the purple blossoms reminds me of grape Kool-Aid, served in the back yard on hot summer days, just as Jacob's ladder begins to bloom.

PLANTING

Seeding: Start seeds in cold frame in spring or fall.

Planting out: Spring.

Spacing: About 12" (30 cm) apart.

GROWING

Grow these plants in **full sun or partial shade**. Tall species grow in **fertile, moist, well-drained** soil. Low-growing species grow in **gravely, very well-drained** soil. Use all types of Jacob's ladder in borders and woodland gardens. Low-growers may also be included in rock gardens and to edge paths. Jacob's ladder seldom required dividing, but you may divide in late summer if necessary. This plant reseeds readily.

P. carneum

RECOMMENDED

P. caeruleum is a tall species, growing 12–36" (30–90 cm) tall and 12" (30 cm) wide. The flowers are lavender blue or sometimes white.

P. carneum is a low-growing, Pacific coast native. It grows 4–16" (10–40 cm) tall and 8" (20 cm) wide. The flowers are pale pink, yellow, light or dark purple. This species is more difficult to grow and requires a sandy soil.

PROBLEMS & PESTS

Jacob's ladder is usually pest free, but may have some problems with powdery mildew. The leaves will scorch in hot afternoon sun.

GARDENING TIPS

Jacob's ladder will bloom for longer periods if regularly deadheaded.

This plant will grow well in containers when used as tall focal points in the center of a large urn or barrel.

P. caeruleum

Japanese Anemone
Windflower
Anemone x *hybrida*

Flower color: Pink, red, purple, blue, yellow or white.
Height: 48–60" (120–150 cm) tall. **Spread:** May spread indefinitely.
Blooms: Late summer and fall.

Anemones are a large and varied group of buttercup relatives. They are valued both as one of the most beautiful and earliest spring flowers and as one of the last plants to continue flowering in late fall. Japanese Anemones are late bloomers appearing at the end of summer when other flowers are fading.

Then name of the Anemone comes from the Greek anemos, meaning 'wind.' The plant grows on windswept mountainsides, which is how it got its name.

PLANTING

Seeding: Slow growing; some species take up to two years to germinate.

Planting out: Spring or fall.

Spacing: 24" (60 cm) apart.

GROWING

Japanese Anemone will grow equally well in **sun or partial shade**. Soil should be **humus-rich, moist and well drained**. Mass planted, they look magnificent in a border in fall. Divide them in early spring or late fall and grow them in containers for a year before planting them back in the garden in spring.

RECOMMENDED

'Max Vogel' has light pink flowers.

'Pamina' has pinkish-red, double flowers.

'White Giant' has white, semi-double flowers.

ALTERNATE SPECIES

A. blanda (Grecian Windflower) is a beautiful spring-flowering species. It grows about 6" (15 cm) tall with an equal spread. The flowers of this species are blue, but there are cultivars available in white and shades of pink and blue. Grow it in full sun in light, well-drained soil. It prefers to dry out during summer while it goes dormant. This species is often sold in fall with tulips and daffodils as a spring-blooming bulb.

A. sylvestris (Snowdrop Anemone) grows to be 14" (35 cm) tall and does well in woodland settings. It is white with a yellow center.

White Giant

GARDENING TIPS

Plant Japanese Anemone behind shrubby roses to support the tall stems. It is slow to establish itself, but then it takes off later in the season.

PROBLEMS & PESTS

Problems with leaf gall, downy and powdery mildew, smut, leaf spot, rust and nematodes are quite common. Less of a problem, but occasionally troublesome, are viruses, caterpillars, slugs and flea beetles.

Japanese Rush
Sweet Flag; Grassy-Leafed Sweet Flag
Acorus gramineus

Flower color: Inconspicuous. The plant is grown as a foliage plant. **Height:** 3–14" (8–35 cm) tall. **Spread:** 4–6" (10–15 cm) wide.

Japanese Rush is a water-loving plant with wide grass-like leaves. The species that are most popular are native to Japan and Northern Asia. Japanese Rush is available in large or miniature varieties. All varieties will grow best in wet or boggy locations. It is a popular choice along the edges of moats around European castles.

PLANTING

Seeding: Not recommended.

Planting out: Spring.

Spacing: About 5" (13 cm) apart.

When old books refer to floors strewn with sweet-smelling rushes it is this plant that is being referred to. Floors littered with leaves are no longer as popular as they once were, but you may still want to try using Japanese Rush leaves for a scented doormat.

GROWING

Japanese Rush will tolerate **full sun or partial shade**. Japanese Rush doesn't just tolerate wet conditions, it requires them. It will even grow when partially submerged in the water at the edge of a pond. This plant is considered an **aquatic** and is one of the smaller pondside plants. It is an excellent choice in a boggy or wet area of the garden or, of course, around the moat of your castle. Divide every three or four years in spring. Pot the divisions and plant them out once they are established.

RECOMMENDED

'Ogon' is 10" (25 cm) tall and 4–6" (10–15 cm) wide with glossy pale green and cream variegated leaves.

'Pusillus' (Dwarf Japanese Rush) is 1½–6" (4–15 cm) tall and 4–6" (10–15 cm) wide. It is a compact plant with glossy, dark green leaves.

'Variegatus' (Variegated Japanese Rush) is 10" (25 cm) tall and 6" (15 cm) wide. The leaves are green with cream stripes.

ALTERNATE SPECIES

A. calamus **'Variegatus'** (Variegated Sweet Flag) is 60" (150 cm) tall and 24" (60 cm) wide. This is a similar, but much larger plant than *A. gramineus* and likes to grow in the same conditions.

GARDENING TIPS

The smaller varieties of Japanese Rush will do nicely in a barrel water garden. The plants may be grown in a submerged water basket sitting on a brick. The plant base should be no more than 4" (10 cm) underwater. *A. gramineus* cultivars suffer in hard freezes and may require winter protection. If planted in containers, they can be stored in a garage or porch for the winter.

When crushed the leaves of Japanese Rush give off a spicy cinnamon scent.

Lady's Mantle

Alchemilla mollis

Flower color: Yellow.
Height: 18" (45 cm) tall. **Spread:** 12–24" (30–60 cm) wide.
Blooms: Early summer to early fall.

This plant is treasured as much for its foliage as for its flowers. The soft green leaves are almost completely round with scalloped edges. When protected from the sun in very hot locations they will continue to look fresh and green all summer. Delicate sprays of greenish-yellow flowers veil the leaves when the plant is in flower. This plant has a very soft appearance and is soothing to the eye when used as a break in a brightly colored border. The slightly cupped leaves collect dewdrops that glisten like diamonds.

The flowers can be enjoyed in fresh or dried arrangements and the leaves can be boiled to make a green dye for wool.

PLANTING

Seeding: Start in early spring and transplant early. Plants may not flower until second year.

Planting out: Spring.

Spacing: 18" (45 cm) apart. Established plants are unlikely to need dividing, but self-seeded seedlings may be moved to other locations if desired.

GROWING

Lady's Mantle will grow in **full sun or light shade**. It does not like too much heat, however, and will look its best if protected from direct sun in hot locations. It prefers to grow in a **rich, moist soil** amended with lots of organic matter and is drought-resistant once established.

Lady's Mantle is ideal for grouping under trees in woodland gardens and along border edges where it softens the often bright colors of other plants. A perfect location is along a pathway that winds through a lightly wooded area. When division is necessary it should be done in early spring before flowering begins.

ALTERNATE SPECIES

A. alpina is a popular small version of Lady's Mantle. It grows to be 3–5" (8–13 cm) tall and up to 20" (50 cm) wide.

GARDENING TIPS

If Lady's Mantle begins to look tired and heat-stressed during summer, it will rejuvenate if cut back lightly. If self-seeding is not desired then the plant should be deadheaded. This can encourage a second flush of flowers.

A. mollis

Poets and alchemists were inspired by the crystal-like dew that collects on the leaves, which was reputed to have magical and medicinal qualities.

A. alpina

A. mollis

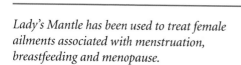

Lady's Mantle has been used to treat female ailments associated with menstruation, breastfeeding and menopause.

Lamium
Spotted Dead-Nettle
Lamium maculatum

Flower color: White, pink or red-purple.
Height: 8" (20 cm) tall. **Spread:** 36" (90 cm) wide.
Blooms: Spring and summer.

Lamium is a low-growing, shade-loving foliage plant with the added attraction of bearing small flowers in spring and early summer. As a groundcover, Lamium will be evergreen most winters and is used in a shaded section of my garden to solve my mole problem. Rich, moist soil full of earthworms was so attractive to the moles that I finally gave up growing grass and covered the ground with easy-going Lamium 'Beacon Silver.' Although the moles are still at work, the lovely Lamium covers the holes and burrows, leaving nothing but an attractive mass of cool green and white foliage. Much nicer and less work than a lawn full of mole holes.

PLANTING

Seeding: Start seeds in spring or fall. Some cultivars may not come true to type from seed.

Planting out: Spring.

Spacing: 12–24" (30–60 cm) apart.

GROWING

Lamium prefers a location that is only **partially shaded** but will tolerate full shade, becoming a bit leggy in deeper shade. Soil should be **average, with plenty of humus** worked in. Lamium prefers to be evenly moist but will tolerate drought as long as it is not in the sun.

A good groundcover for woodland and shade gardens and under shrubs in the border, Lamium is likely to keep the weeds down. If allowed to spread unchecked, Lamium can be troublesome itself—keep it away from less vigorous plants as it may smother them. Divide in fall.

The species name maculatum *refers to the spots that appear on the leaves. The variation in leaf markings is often the basis for a new cultivar. Keep your eyes open for unusual markings on your plants and you may wind up with a cultivar you can name after yourself.*

'Beacon Silver'

RECOMMENDED

'**Aureum**' has gold-colored leaves with white stripes down the centers and pink flowers.

'**Beacon Silver**' has green-edged silver leaves and pink flowers.

'**Chequers**' has green leaves with silver stripes down the centers and mauve flowers.

'**White Nancy**' has green-edged, silver leaves and white flowers.

ALTERNATE SPECIES

L. galeobolon '**Variegatum**' is similar in habit and size to *L. maculatum* but has yellow flowers.

GARDENING TIPS

Lamium will remain more compact if sheared back after flowering. The more fertile the soil the more vigorously the plant will grow. If the plant is becoming invasive, pull some of it up making sure to remove the fleshy roots.

Bare spots in the plant are usually the result of the plant drying out too often. Divide and replant if bare spots become unsightly.

PROBLEMS & PESTS

Slugs and snails are the most common problems. Powdery mildew, downy mildew and leaf spot problems are possible but infrequent.

Lamium is also known as Dead-nettle because its leaves look like stinging nettle.

'Variegatum'

Lewisia
Bitterroot
Lewisia spp.

Flower color: Pink, magenta, white, purple, yellow or orange.
Height: 6–8" (15–20 cm) tall. **Spread:** 6–12" (15–30 cm) wide.
Blooms: Spring and early summer.

The thick, fleshy leaves arranged in a whorl around a central bud are like a bull's-eye that targets the spectacular wand of bi-colored blooms that appear in May. It is a sight worth waiting for all year long and features tiny flowers you can bring indoors to enjoy up close. You may want to put in an alpine or scree garden or raised-bed rock garden just to have an excuse to grow lewisia.

PLANTING

Seeding: Start seeds in cold frame in fall.

Planting out: Spring.

Spacing: 6–12" (15–30 cm) apart.

GROWING

Grow lewisia in **full sun**. Soil should be **moderately fertile, humus-rich, neutral or acidic and very well drained**. This is a perfect plant for dry, exposed areas such as in the crevices of a rock wall, rock garden or dry bank. Lewisia does not need dividing. Allow lewisia to dry out after blooming—it may require protection from heavy rain.

RECOMMENDED

L. columbiana is native along the coast from British Columbia to Oregon. It grows about 6" (15 cm) tall and wide and bears magenta or light pink flowers.

L. cotyledon var. *howellii* is native to northwestern California. It grows up to 12" (30 cm) tall and spreads up to 10" (25 cm) wide. The flowers are light pink with dark pink veins. The leaves have undulating margins (photo on opposite page).

L. tweedyi is native to the Pacific Northwest. It grows up to 8" (20 cm) tall and 12" (30 cm) wide, forming a clump of rosettes. The flowers are white, yellow, apricot or peach pink.

GARDENING TIPS

The small offsets that develop at the base of the plant may be separated in early summer and rooted in pots. Grow the young plants in pots for a year before planting them in the garden. Mulch with gravel to keep excess moisture away from the fleshy leaves.

PROBLEMS & PESTS

Occasional problems with snails, slugs, aphids, rust or stem rot are possible.

L. tweedyi

All the species of this plant originated in western North America. The genus was named after Captain Meriwether Lewis of the famous Lewis and Clark expeditions.

L. columbiana

Lily-of-the-Valley

Convallaria majalis

Flower color: White is most common, but pink varieties are also available.
Height: 9" (23 cm) tall. **Spread:** May spread indefinitely.
Blooms: Spring.

The fragrance that wafts from the small, white, bell-shaped flowers is so rich and enchanting that it was often used as the base for fine French perfume. Gardeners should enjoy this classic spring bloomer by snipping a sprig and tucking it into a button hole or the cuff of a gardening glove.

PLANTING

Seeding: Remove flesh from ripe seeds and plant in cold frame. Seeds ripen by mid- to late summer.

Planting out: Spring or fall.

Spacing: 9–12" (23–30 cm) apart. They will quickly fill an area.

GROWING

Lily-of-the-valley will grow in **full sun or full shade**. Soil should be **humus-rich and moist**, but this plant will tolerate dry conditions.

This versatile groundcover grows in a variety of locations. It makes a beautiful plant for naturalized woodland gardens, perhaps bordering a pathway or underneath shade trees where little else, including grass, will grow. It also makes a good groundcover in a shrub border, where the dense growth will keep the weeds down but won't interfere with the shrubs.

Lily-of-the-valley is currently being researched for its potential medicinal qualities.

Divide as often as desired. Leaves grow from small pips, or eyes, that run along the root. Divide the root leaving at least one pip on each root. Lily-of-the-valley may be left without dividing for several years.

RECOMMENDED

'Flore Pleno' has white, double flowers. It tends not to be as invasive as other varieties.

'Fortin's Giant' has larger leaves and flowers than other varieties. It grows up to 12" (30 cm) in height.

'Rosea' is a slightly smaller plant with light pink flowers. It grows up to 8" (20 cm) tall.

European legend claims the origin of Lily-of-the-valley to be either from the tears of the Virgin Mary shed at the cross or the tears of Mary Magdalen shed at Christ's tomb.

GARDENING TIPS

Lily-of-the-valley can be quite invasive. It is a good idea not to locate it with plants that are likely to be overwhelmed, such as those in alpine rock-garden plants. Give Lily-of-the-valley plenty of space to grow and it will do fine. Avoid planting it where you will later have to spend all your time trying to remove it.

Lily-of-the-valley is well known for the delightful scent of its flowers. In fall, dig up a few roots and plant them into pots. Keep the pots in a sheltered section of the garden for winter. In spring you can bring the pots indoors to enjoy both the flowers and their scent.

PROBLEMS & PESTS

Molds and stem rot are the most likely problems to be encountered.

Lungwort

Pulmonaria spp.

Flower color: Blue, red, pink or white.
Height: 12" (30 cm) tall. **Spread:** 24" (60 cm) wide.
Blooms: Spring.

These attractive, shade-loving groundcover plants are often grown for their attractive foliage alone. The leaves can be green with white or silver spots or completely silver. The ugly name given this beautiful plant comes from the shape of the leaves that are much like the lobes of a lung—pointed at the ends but broad across the middle.

Pulmonaria is a traditional culinary and medicinal herb. The young leaves may be added to soups and stews to flavor and thicken the broth. The spotted leaves also make an attractive addition, when dried, to potpourri.

PLANTING

Seeding: Don't always come true from seed. Start freshly ripened seeds in containers outdoors.

Planting out: Spring.

Spacing: 24" (60 cm) apart.

GROWING

Lungwort prefers to grow in **full or partial shade**. Soil should be **fertile, humus-rich and moist, but not soggy**. It is a useful groundcover plant in shady gardens, woodlands, shrub borders, ponds or streamsides. Divide after flowering or in fall.

RECOMMENDED

'Janet Fisk' has densely silver-spotted leaves and pink flowers that turn blue as they age.

'Mrs. Moon' has pink buds that open to light mauve-blue.

'Pink Dawn' has dark pink flowers that turn purple as they age.

'Roy Davidson' has sky blue flowers that bloom in early spring.

'Sissinghurst White' has white-spotted leaves and pink buds that open to white flowers.

GARDENING TIPS

Shear back the foliage after blooming to keep the plants tidy.

PROBLEMS & PESTS

Lungwort may get powdery mildew if the ground dries out for extended periods. Remove and destroy damaged leaves. Slugs and snails may attack the young leaves.

There are over twenty common names for this plant. Many are after biblical characters such as Abraham, Isaac and Jacob, Adam and Eve, Children of Israel, and Virgin Mary.

Lupine

Lupinus spp.

Flower color: White, cream, yellow, pink, red, orange,
purple or blue and often bicolored.
Height: 30–36" (75–90 cm) tall. **Spread:** 12–16" (30–40 cm) wide.
Blooms: Spring and summer.

A field of lupine, splashing across mountainsides, is a wilderness hiking experience not soon forgotten with the flower colors blending from one shade to another like melting ice cream. The flower stalks are shorter and wider than the more refined Delphinium or casual and playful Foxglove when used as tall bloomers at the back of the perennial border. The shape and texture added by a stand of lupine is as unique as the shades of the flowers.

PLANTING

Seeding: Soak seeds in warm water for 24 hours then plant directly outdoors in late fall or early spring. If you are starting seeds indoors you may need to place planted seeds in refrigerator for four to six weeks after soaking them.

Planting out: Spring or fall.

Spacing: 12" (30 cm) apart.

GROWING

Grow these plants in **full sun or partial shade**. Soil should be **average to fairly fertile, sandy, well drained and slightly acidic**. Protect plants from drying winds. Lupines are wonderful when massed together in borders or in cottage or natural gardens. Lupines don't like to have their roots disturbed. The small offsets that develop at the base of the plant may be carefully separated and replanted to propagate the plant.

Lupines are in the same plant family as beans and peas. However, the pods and seeds of lupine will cause stomach upset if ingested.

Russell hybrids

RECOMMENDED

Russell hybrids (Russell Lupines) are a group of hybrids developed from crossing several species of Lupine. The result is this popular dwarf group. They grow about 24–36" (60–90 cm) tall and 12–18" (30–45 cm) wide. They bear flowers in the widest range of solid colors and bi-colors.

L. latifolius is a Pacific coast native found from Washington to California. It grows up to 48" (120 cm) tall and 24" (60 cm) wide. Spikes of blue or purple flowers bloom in early summer.

L. polyphyllus (Washington Lupine) is a Pacific native from British Columbia to California. It grows 24–60" (60–150 cm) tall and 24" (60 cm) wide. The flowers, in colors of blue, white or pink, bloom throughout summer.

GARDENING TIPS

Deadheading is generally recommended for lupines to encourage more flower spikes later in the season. On the other hand, lupines may self-seed if the spent spikes are left in place. Leave a few spikes in place after the flowers are gone if you want to have some new plants grow in to replace the older ones that die out.

PROBLEMS & PESTS

Aphids are the biggest problem. Slugs, snails, fungal and bacterial spots, downy and powdery mildew, rust, stem rot and damping off are all possible problems.

The fuzzy, peapod-like capsules that form along the spike once the flowers fade can be removed or left to ripen so the seeds can be collected.

Mallow

Malva spp.

Flower color: Purple, pink, white or blue.
Height: 36–48" (90–120 cm) tall. **Spread:** 24" (60 cm) wide.
Blooms: Summer and fall.

Upright and bursting with cottage garden charm, mallows don't mind rocky soil and will bloom happily against a fence or wall looking like friendly Hollyhocks, but with soft, gray-green leaves that are the perfect background for pastel flowers. To add to the storybook charm, friendly, easygoing mallows invite butterflies and hummingbirds to visit the garden often.

PLANTING

Seeding: Direct sow in garden in early spring or early summer.

Planting out: Spring or summer.

Spacing: 24" (60 cm) apart.

GROWING

Grow mallow in **full or partial sun.** Soil should be **average to moderately fertile, moist and well drained.** Mallow is drought-tolerant. In very rich soils the plants may require staking. Use mallow in a mixed border or in a wild or cottage garden. Deadhead the flowers to keep the plant blooming until October. Mallow will reseed readily; it is short-lived and will not require dividing.

Mallow is reputed to have a calming effect when ingested and was used in the Middle Ages as an antidote for aphrodisiacs and love potions.

M. moschata

The common name of M. sylvestris, Cheeses, refers to the unripe seed capsules. Although they taste nothing like cheese they are edible and can be added to salads.

RECOMMENDED

M. alcea (Hollyhock Mallow) has 2–3" (5–8 cm) wide flowers in shades of pink from early summer to early fall. The plant may be quite erect.

M. moschata (Musk Mallow) has large, over 3" (8 cm) wide, pale pink or white flowers throughout summer. The plant is erect, growing to 36" (90 cm) tall, and bushy.

M. sylvestris (Cheeses) has dark pink flowers with purple veins. It can be a spreading or upright species. This plant has several cultivars, many with stunning flowers. **'Primley Blue'** is a prostrate cultivar growing only 8" (20 cm) tall and spreading 12–24" (30–60 cm) along the ground. **'Braveheart'**

'Primley Blue'

is an upright cultivar growing up to 36" (90 cm) tall (photo on p. 226). **'Bibor Felho'** is also upright growing up to 6' (2 m).

GARDENING TIPS

Mallow may be propagated with basal cuttings taken in spring or by self-seeding. Mallow can be used as a cut flower. Light pruning encourages new growth and more flowers. Cut the plants back in May by one-half to make them more compact and bushy.

PROBLEMS & PESTS

Problems with rust and leaf spot are possible.

'Primley Blue'

'Bibor Felho'

Marsh Marigold
Cowslip
Caltha palustris

Flower color: Yellow or white.
Height: 4–24" (10–60 cm) tall. **Spread:** 10–30" (25–75 cm) wide.
Blooms: Spring.

This plant is called Cowslip by dairy farmers because of the way it grows along stream and creek beds, causing hoofed feet to slip along the banks. Although it may not welcome cows to the garden, it is a wonderful way to welcome spring in any boggy, wet spot.

PLANTING

Seeding: Sow seeds in moist soil in summer. Unlikely to sprout before spring, so mark locations clearly.

Planting out: Spring.

Spacing: 10–30" (25–75 cm) apart.

GROWING

Marsh Marigold will grow in **full sun or partial shade**. Soil should be **constantly moist or wet**. Marsh Marigold will even grow in water as deep as 6" (15 cm). Marsh Marigold is an ideal plant to include in a water, stream or bog garden. It grows and flowers quickly in spring then dies back after flowering has finished. It is an excellent plant to use with plants that are slow in spring to fill in but later in summer require more room. Marsh Marigold works well in naturalized gardens. Divide and replant in early summer, after flowering is complete, every two or three years.

RECOMMENDED

The straight species is one of the most beautiful of wet garden plants, but the following options are also attractive.

'Alba' has white flowers and is less vigorous than the straight species. It grows 9" (23 cm) tall and 12" (30 cm) wide.

'Flore Pleno' has yellow, double flowers. It grows 10" (25 cm) tall with equal width.

C. p. var. palustris (Giant Marsh Marigold) is native to the Pacific Northwest and has yellow flowers. It is much larger than the other varieties, growing 24" (60 cm) tall and 28–30" (70–75 cm) wide.

GARDENING TIPS

Even if you aren't interested in having a water garden you can still grow Marsh Marigold. You can grow this plant in the wettest part of your garden or even in a shady border as long as you are willing to keep it very well watered.

PROBLEMS & PESTS

Marsh Marigold is susceptible to powdery mildew and rust, but if the plant is kept well watered and stress-free these problems can be avoided.

The species name, palustris, refers to this plant's growing in wet, boggy areas.

Meadow Rue

Thalictrum spp.

Flower color: Pink, yellow or purple.
Height: 36–60" (90–150 cm) tall. **Spread:** 12–24" (30–60 cm) wide.
Blooms: Spring or summer.

Airy, delicate and soft-looking foliage makes this an appealing plant to use with more broadly textured perennials. The clusters of flowers are an extra bonus, easy to cut for lacy indoor bouquets or to dry and enjoy all winter long. The Dusty Meadow Rue has gray-green foliage that brings out the color of pink flowers or companion plants with yellow leaves.

PLANTING

Seeding: Start seeds in containers in cold frame in fall or spring.

Planting out: Spring.

Spacing: 15–28" (38–70 cm) apart.

GROWING

Grow these plants in **full sun or partial shade**. Soil should be **humus-rich, moist and well drained**. In the middle or at the back of a border, meadow rue makes a soft backdrop for bolder plants and flowers. Some of the smaller species are useful in a rock garden and all are beautiful when naturalized in an open woodland or meadow garden.

These plants rarely need to be divided. If necessary, divide in fall or in spring when the foliage begins to develop. They may take a while to re-establish themselves once they have been divided or have had their roots disturbed.

T. aquilegifolium

Meadow rue is a late-starting plant that will fill in the space left by spring plants such as Marsh Marigold, which go dormant in the summer.

RECOMMENDED

T. aquilegiifolium (Columbine Meadow Rue) has pink or white plumes of flowers. The leaves are similar in appearance to those of the columbine. It grows up to 36" (90 cm) tall and 18" (45 cm) wide.

T. delvayi 'Hewitt's Double' is a late summer and fall bloomer growing 48" (120 cm) or taller and spreading 24" (60 cm). The tiny purple flowers are pom-pom-like and very numerous.

T. minus 'Adiantifolium' is 36" (90 cm) tall and 24" (60 cm) wide with very delicate, fern-like foliage that looks a bit like that of the maidenhair fern. The flowers are white and held in loose clusters. The flowers are often used dry or fresh in flower arrangements.

T. rochebrunianum 'Lavender Mist' is 60" (150 cm) tall and 12–24" (30–60 cm) wide. It forms clumps with delicate foliage and lavender purple blooms.

'Lavender Mist'

GARDENING TIPS

These plants often do not emerge until quite late in spring. Mark the location where they are planted so that you do not inadvertently disturb the roots if you are cultivating their bed before they begin to grow.

Do not plant individual plants close together as their stems will become tangled.

PROBLEMS & PESTS

Infrequent problems with powdery mildew, rust, smut, leaf spots and slugs. Deer resistant.

'Hewitt's Double'

Meadowsweet
Filipendula spp.

Flower color: White, light pink or deep rose pink.
Height: 24–96" (60–245 cm) tall. **Spread:** 18–48" (45–120 cm) wide.
Blooms: Late spring and summer.

Add this perennial if you would like some history in your garden. Meadowsweet has been used for generations to treat such ailments as kidney stones, and gathering the flower clusters from fields and alongside streams and tossing the blooms onto the dusty road was the traditional way for the peasants to welcome Queen Elizabeth I to their villages. For this reason, some species of meadowsweet are called Queen of the Prairie or Queen of the Meadow.

PLANTING

Seeding: Start in cold frame in fall for germination in spring. Keep soil evenly moist over winter to encourage even germination.

Planting out: Spring.

Spacing: 18–36" (45–90 cm) apart.

GROWING

Filipendula will grow equally well in **full sun or partial shade** and will prefer partial shade if the soil dries out too often. Soil should be **deep, fertile and moist**. It particularly likes leaf compost mixed into the soil.

This is an excellent plant for bog gardens or wet sites. Grow *Filipendula* alongside streams or in moist meadows. It may also be grown in the back of a border, as long as it is kept well watered. Divide in spring or fall.

F. ulmaria was a popular flavoring for ales and meads in medieval times. It is thought that the name meadowsweet is derived from the Anglo-Saxon word medesweete *because it was often used to flavor mead.*

F. ulmaria

RECOMMENDED

F. palmata (Siberian Meadowsweet) has pale or deep pink flowers that often fade to white with age. The leaves have soft hairs covering their undersides. This species grows up to 48" (120 cm) tall and 24" (60 cm) wide.

F. rubra (Queen of the Prairie) grows to be 72–96" (180–245 cm) tall and 48" (120 cm) wide. **'Venusta'** has deep flowers on red stems.

F. ulmaria (Meadowsweet; Queen of the Meadow) has cream-colored flowers in large clusters. It grows to be 24–36" (60–90 cm) tall and 24" (60 cm) wide. In past times it was used to flavor mead and ale. It is gaining popularity today when used to flavor vinegars and jams. It may also be made into a pleasant wine (much the same way as dandelion wine is made).

F. vulgaris (Dropwort) is a low-growing species that prefers dry soil. It is a good choice if you have a dry area of the garden. It grows up to 24" (60 cm) tall and 18" (45 cm) wide. **'Rosea'** has pink flowers and **'Flore Pleno'** has white, double flowers.

'Flore Pleno'

GARDENING TIPS

These plants may be deadheaded if desired, but the faded seedheads are quite attractive when left in place.

PROBLEMS & PESTS

Powdery mildew, rust and leaf spot are problems.

In the 16th century it was customary to strew floors with rushes and herbs to warm the floor underfoot, to freshen the air and combat infections; meadow-sweet was the favorite choice of Queen Elizabeth I.

F. rubra

Michaelmas Daisy
Aster spp.

Flower color: Red, white and shades of purple and pink.
Height: 48–60" (120–150 cm) tall. **Spread:** 24–36" (60–90 cm) wide.
Blooms: Summer and fall.

Asters are hardworking plants that offer fresh-as-a-daisy blooms when the dog days of summer have beat most other perennials into submission. Add them to the garden for their colorful company and you will also have birds, butterflies and bees arriving for a late summer or fall garden party.

PLANTING

Seeding: Start seeds in cold frame in spring or fall.

Planting out: Spring or fall.

Spacing: 24–36" (60–90 cm) apart.

GROWING

Grow these plants in **full sun or partial shade**. Soil should be **fertile, moist and well drained**. Attractive plants to include in the middle of the border, they also look good in naturalized or wild gardens. Divide every two or three years in spring or fall. Plants will decline rapidly if not divided frequently.

A. *novi-belgii* (upper left) and 'Purple Dome' (lower right)

This old-fashioned flower was once called starwort because of the many petals that radiate out from the center.

What looks like a single flower on a Michaelmas daisy, or other daisy-like plants, is actually a cluster of many tiny flowers. Look closely at the center of the flower-head and you will see all the tiny individual flowers.

RECOMMENDED

A. x *frikartii* '**Monch**' grows to be 30" (75 cm) tall with an open branching habit. It has one of the longest blooming periods of the asters and has lavender purple flowers.

A. novae-angliae (New England Aster) is the parent plant of many popular and attractive cultivars. Flowers are white, dark purple and salmon, rose or light pink. '**Purple Dome**' is a compact, spreading plant growing to 18" (45 cm) tall and 30" (75 cm) wide. The flowers are semi-double, deep purple and they completely cover the plant when in bloom. This cultivar is tolerant of wet soil.

A. novi-belgii (New York Aster) is another parent of popular cultivars. Reds, blues, purples, pinks and whites are all available in the cultivars. The plants are larger than the New England Aster. Large cultivars are often crosses between the New York and New England Asters. *A.* x *pringlei* '**Monte Casino**' has yellow stems with clouds of white flowers with yellow centres. The flowers continue after frost.

A. novae-angliae

GARDENING TIPS

To make the plants bushier and more compact to avoid staking, cut them back with hedge shears in early spring and again in late spring.

PROBLEMS & PESTS

Some fungal problems as well as trouble with aphids, slugs, snails, rosy blister gall and nematodes are possible. Deer resistant.

A. novi-belgii

Monkshood
Wolfsbane
Aconitum napellus

Flower color: Purple, blue or white.
Height: 60" (150 cm) tall. **Spread:** 12" (30 cm) wide.
Blooms: Late summer.

Monkshood is a beautiful old favorite that thrives in damp and shady conditions. All parts of the plant are poisonous to pets and humans. Its use in the garden will deter wildlife and local cats from making a home in your flowerbeds. The possible dangers of Monkshood shouldn't deter you from enjoying its beautiful flowers. This plant is a lovely addition to a damp corner of the garden.

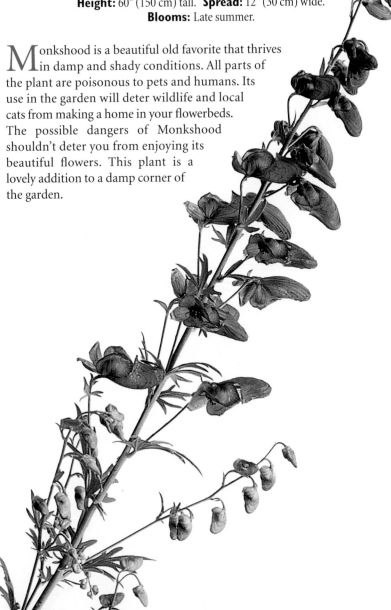

PLANTING

Seeding: Germination may be irregular. Seeds sown in garden in spring may bloom in following summer; seeds planted later will not be likely to bloom for yet another year.

Planting out: Spring. Bare-rooted tubers may be planted in fall.

Spacing: 18" (45 cm) apart.

GROWING

Monkshood prefers to grow in a **slightly shaded area** but will tolerate sun if the climate is cool. It will grow in any **moist** soil but prefers to be in a **rich** soil with lots of organic matter worked in. Monkshood prefers conditions to be on the cool side. It will do poorly when the weather gets hot, particularly if conditions do not cool down at night. Mulch the roots to keep them cool. This plant requires a period of dormancy in winter.

The alternate common name, Wolfsbane, refers to the former use of this plant as a wolf poison.

Monkshood is the perfect plant to grow in cool, boggy locations along streams or next to ponds. It makes a tall and elegant addition to a woodland garden in combination with lower growing plants. Monkshood prefers not to be divided as it may be a bit slow to re-establish itself. If division is desired to increase the number of plants then it should be done in late fall or early spring.

RECOMMENDED

A. x *bicolor* has royal blue and white flowers.

'**Sparks Variety**' has late and long-blooming, tall, wiry candelabras of navy blue flowers.

ALTERNATE SPECIES

A. carmichaelii '**Arsendii**' is late flowering and does not require staking.

A. septentrionale '**Ivorine**' is low growing, to 24" (60 cm), with white flowers.

A. x bicolor

GARDENING TIPS

When dividing or transplanting Monkshood, the crown of the plant should never be planted at a depth lower than where it was previously growing. Burying the crown any deeper will only cause it to rot and the plant to die.

Do not plant near tree roots because it cannot handle the competition.

Monkshood contains one of the most toxic plant compounds ever discovered. The ancient Chinese and the Arabs used the juice of Monkshood to poison arrow tips.

Mullein
Verbascum spp.

Flower color: Yellow or white.
Height: 36–72" (90–180 cm) tall.
Spread: 18–24" (45–60 cm) wide.
Blooms: Summer.

The broad, velvety gray leaves call out to be petted, and the tall lines of a blooming mullein make a great architectural feature. Throughout history, soft mullein foliage has been used for bandages and compresses, and they still offer first aid to any garden with dry, poor soil that needs a shot of ghostly color, cloud-scraping drama and pleasing texture.

PLANTING

Seeding: Start seeds in cold frame in late spring or early summer for flowers in second season.

Planting out: Spring.

Spacing: 24" (60 cm) apart.

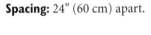

GROWING

Grow these plants in **full sun**. Soil should be **poor, alkaline and well drained**. When planted in soil that is too rich the plants become floppy and fall over. Use in the middle or at the back of a sunny border. Mullein is also good in natural gardens. Mullein may be divided in spring, though this is rarely necessary.

RECOMMENDED

V. chaixii (Nettle-leaved Mullein) grows up to 36" (90 cm) tall and bears yellow flowers. The cultivar **'Album'** has white flowers with mauve centers.

V. nigrum (Dark Mullein) has dark yellow flowers with purple centers (photo on opposite page).

V. olympicum (Olympic Mullein) has very woolly, silver gray leaves and bright yellow flowers. It grows up to 72" (180 cm) tall.

V. phoenicium (Flush of White; Purple Mullein) is 24" (60 cm) tall with loose spikes of white, pink or purple flowers. There are several popular hybrids of this species crossed with one of the yellow-flowered species. **'Helen Johnson'** has orange flowers. **'Mont Blanc'** has white flowers.

GARDENING TIPS

Removing the spent flower spikes may encourage the plants to bloom again. The plants may self-seed. The plants are short-lived, but self-seeding will keep them growing in your garden.

Mullein will rot if planted in a location with moist soil or where water from run-off or overflowing gutters hits the foliage.

PROBLEMS & PESTS

Powdery mildew, fungal leaf spots and caterpillars may cause problems. Deer and slug resistant.

Some species of mullein are considered weeds. Each capsule on a flower spike can produce hundreds of seeds.

V. olympicum

'Helen Johnson'

Oriental Poppy

Papaver orientale

Flower color: Red, orange, pink or white.
Height: 12–36" (30–90 cm) tall. **Spread:** 12–36" (30–90 cm) wide.
Blooms: Spring and early summer.

This is one perennial to enjoy even before it blooms. Gigantic, nodding, fuzzy buds crack open slowly to unfurl delicate tissue-paper petals that wave and dance on the top of wiry stems. Even the seedpods are attractive, resembling pepper shakers ready to spill forth the tiny black seeds from dried brown capsules.

PLANTING

Seeding: Start seeds in cold frame in spring.

Planting out: Spring.

Spacing: 24" (60 cm) apart.

GROWING

Grow these plants in a location that receives **full sun.** Soil should be **average to fertile and must be well drained.** Small groups of Oriental Poppy look attractive in an early summer border. Division is rarely required and may be done in fall once the new rosettes begin to form. Plants will die back after flowering and send up fresh new growth in late summer, which should be left in place for winter insulation.

RECOMMENDED

'Allegro' has bright scarlet red flowers.

'Black and White' has white flowers with black markings at the bases of the petals.

ALTERNATE SPECIES

P. nudicaule **'Garden Gnome'** is 12" (30 cm) tall with bright yellow or orange blooms.

GARDENING TIPS

Plant Oriental Poppy with plants that get bushy as summer wears on.

Because it goes completely dormant by midsummer, Oriental Poppy may leave bare spots in a border. Baby's Breath and Catmint make good companions and will fill any blank spots left in the border later in summer.

When the foliage turns yellow after blooming, cut Oriental Poppy to the ground or it will attract slugs.

PROBLEMS & PESTS

Powdery mildew, leaf smut, gray mold, root rot and damping off may occur. Deer resistant.

Use of poppy seeds in cooking and baking can be traced as far back as to the ancient Egyptians.

Pearly Everlasting

Anaphalis margaritacea

Flower color: White with yellow eyes.
Height: 8–36" (20–90 cm) tall. **Spread:** 6–24" (15–60 cm) wide.
Blooms: Mid- to late summer.

Silver-leaved plants are always popular for their ability to both bring out the colors in other plants and to soothe the eye in a border with a lot of color. This attractive plant is one of the few silver-leaved plants that enjoys a moist location.

PLANTING

Seeding: Start seeds in cold frame in spring. Direct sow in garden in early summer.

Planting out: Spring.

Spacing: 6–24" (15–60 cm) apart.

GROWING

Pearly Everlasting needs a location in **full sun**. Soil should be **average to rich and well drained**. This plant prefers to be watered regularly to prevent the lower leaves from dropping. Pearly Everlasting makes an attractive addition in a variety of situations. It is one of the only silver-leaved plants that may be used as an accent around a water feature. Its late summer clusters of papery white flowers are a good choice for the middle or foreground of a bed or border when other plants are past their best.

This vigorous plant should be divided every three or four years in spring. Pearly Everlasting grows by sending out runners that grow new plants wherever they touch the soil. You can tell when it is time to divide this plant because the center, where the most mature growth is located, will begin to die out.

The clustered strawflower-like blooms make good additions to fresh or dried arrangements.

RECOMMENDED

A. margaritacea is the most drought-tolerant species. **'Cinnamomea'** has broad leaves that have cinnamon-colored undersides and compact flowers.

ALTERNATE SPECIES

A. nepalensis **var.** *monocephala* makes a good rock-garden plant as it only grows 4–8" (10–20 cm) tall and 12" (30 cm) wide.

A. sinisa **ssp.** *morii* is another miniature species that makes a good groundcover for a smaller garden, growing 8" (20 cm) tall and spreading up to 24" (60 cm).

GARDENING TIPS

The plant may be pruned in early to midsummer to promote more blooms and denser growth.

PROBLEMS & PESTS

Caterpillars sometimes attack this plant in early summer. Stem rot, rust and *Septoria* leaf spot are also occasional problems during hot, humid weather.

Penstemon
Beard Tongue
Penstemon spp.

Flower color: White, yellow, light pink, rose pink or purple.
Height: 8–24" (20–60 cm) tall. **Spread:** 10–12" (25–30 cm) wide.
Blooms: Spring and summer.

The colorful stalks of penstemon in the garden always remind me of mountain hikes, and the lilac blue flowers combine well with bright yellow daylilies and the deep purple foliage of *Heuchera* 'Purple Palace.' Grow them with Foxgloves and Snapdragons for vertical lines in the back of the garden bed and include them wherever you want to welcome hummingbirds and butterflies on a summer day.

PLANTING

Seeding: Sow seeds in late winter or spring. Seeds require soil temperature of 55–64° F (13–18° C).

Planting out: Spring or fall.

Spacing: About 12" (30 cm) apart.

GROWING

Grow these plants in a location with **full sun or partial shade**. Soil should be **fertile and well drained**. Penstemons are drought-tolerant. The attractive flowers are lovely to include in a mixed border, cottage garden or rock garden. Divide every two or three years in spring.

RECOMMENDED

There are many popular hybrid cultivars available.

'Alice Hinley' has mauve flowers.

'Apple Blossom' has pink-flushed, white flowers (photo on opposite page).

'Snowstorm' has white flowers.

P. barbatus (Beard Tongue Penstemon) has short hairs on the lower petals of its foxglove-like flowers.

P. barrettiae is native to northwestern North America. It bears purple flowers in early summer.

P. confertus is native to northwestern North America. It bears yellow flowers in early summer.

GARDENING TIPS

To support the tall stems, insert twiggy branches around the plants in spring before they grow tall and floppy. Pinch plants when they are 12" (30 cm) tall to encourage bushy growth.

PROBLEMS & PESTS

Slugs and snails may damage young foliage. Powdery mildew, rust and leaf spots may also be problems.

'Snowstorm'

There are over 200 species of Penstemon found native in varied habitats throughout North and South America.

'Alice Hinley'

Peony

Paeonia spp.

Flower color: White, cream, yellow, pink, red or purple.
Height: 24–36" (60–90 cm) tall, but may be shorter or taller, depending on cultivar.
Spread: The spread is usually equal to the height.
Blooms: Spring and early summer.

Peonies are the work horses of the perennial garden, thriving for generations in the same spot, but gardeners may forget that the wide, deep green leaves that adorn the plants all summer and into fall are almost as impressive as their huge blooms. It is this attractive foliage that can be put to work to shield from the eye the yellowing foliage of spring bulbs past their prime. Peonies can also offer support to taller plants in the back of the border, providing a soft cushion for the leaning stems of Delphiniums or lilies. Skirt the foundations of the house with peonies, or use them as a spring and summer hedge to divide the garden into different areas.

PLANTING

Seeding: Not recommended. Seeds may take two or three years to germinate and many more years to grow to flowering size.

Planting out: Spring. If purchased as bare, tuberous roots, they should be planted in fall, when they are dormant.

Spacing: 24–36" (60–90 cm) apart.

GROWING

Peonies prefer **full sun**, but they will tolerate partial shade. The planting site should be well prepared before the plants are introduced. Peonies like **fertile, humus-rich, moist and well-drained** soil with lots of compost.

These are wonderful plants that look great in a border when combined with other early flowering plants. They may be underplanted with bulbs and other plants that will die down by midsummer. The emerging foliage of peonies will hide the dying foliage of spring plants. Division is not required but is usually the best way to propagate new plants. Division should be done in fall.

In the past peonies were used to cure a variety of ailments. They are named after Paion, who was the physician to the Greek gods.

RECOMMENDED

There are countless cultivars and hybrids available.

'Dawn Pink' has single pink flowers with bright yellow stamens.

'Duchess de Nemours' has fragrant, double, white, yellow-tinged flowers.

'Sarah Bernhardt' was introduced in 1856. It has large, fragrant, double, light pink flowers.

'Shirley Temple' has large, double flowers that open pink and fade to white as they mature.

P. officinalis **'Alba Plena'** has white, double flowers. **'Rubra Plena'** red double flowers.

GARDENING TIPS

Planting depth is a very important factor in determining whether or not a peony will flower. Too shallowly or, more commonly, too deeply planted tubers will not flower. The buds or eyes on the tuber should be 1–2" (2.5–5 cm) below the soil surface.

Cut back the flowers after blooming and remove any blackened leaves to prevent the spread of *Botrytis* blight. Red peonies are more susceptible to disease.

Use wire tomato cages around the plants in early spring to support heavy flowers.

PROBLEMS & PESTS

Peonies may get *Verticillium* wilt, ringspot virus, tip blight, stem rot, *Botrytis* blight, leaf blotch, Japanese beetle or nematodes.

Despite their exotic appearance peonies are tough and hardy perennials that can even survive winter temperatures of -40°F (-40°C).

Phlox

Phlox spp.

Flower color: White, blue, purple or pink.
Height: 2–48" (5–120 cm) tall. **Spread:** 12–36" (30–90 cm) wide.
Blooms: Late spring to fall.

This old-fashioned favorite invokes memories of a gentler time when ladies in hats and gloves took tea in the garden. The taller phlox are majestic and intense in color, the white varieties glow in the moonlight and the groundhugging Creeping Phlox are headliners in a rockery. Their tiny blooms are so numerous that no foliage peeks through the flowers.

PLANTING

Seeding: Start seeds in cold frame in fall or spring.

Planting out: Spring.

Spacing: 12–36" (30–90 cm) apart.

GROWING

Periwinkle and Moss Phlox prefer to grow in **partial shade** and Garden Phlox prefers to grow in **full sun**. All three species like **fertile, humus-rich, moist, well-drained** soil. Low-growing species are useful in a rock garden or at the front of a border. Taller species may be used in the middle of a border where they are particularly effective if planted in groups. Divide in fall or spring.

RECOMMENDED

P. adsurgens (Periwinkle Phlox) is 12" (30 cm) tall and wide, with pink flowers that bloom from late spring to early summer. **'Wagon Wheels'** has narrower petals than the straight species (photo on opposite page). *P. a.* **var.** *alba* has white flowers.

P. paniculata (Garden Phlox) has many cultivars that vary greatly in size from 20–48" (50–120 cm) tall with a spread of 24–36" (60–90 cm) wide. There are many colors available, often with contrasting centers. Garden Phlox blooms from summer to fall.

P. subulata

Phlox come in many forms from low-growing creepers to tall, clump-forming uprights. The many species can be found in varied climates from dry, exposed mountainsides to moist, sheltered woodlands.

P. subulata (Creeping Phlox; Moss Phlox; Moss Pinks) is very low growing, only 2–6" (5–15 cm) tall, with a spread of 20" (50 cm) wide. It also has cultivars with flowers that bloom in various colors from late spring to early summer.

GARDENING TIPS

Periwinkle and Creeping Phlox spread out horizontally as they grow. The stems grow roots where they touch the ground. These plants are easily propagated by detaching the rooted stems in spring or early fall.

Do not prune Creeping Phlox in fall—it is an evergreen and will have next spring's flowers already forming.

Garden Phlox requires good air circulation to prevent mildew.

P. paniculata

PROBLEMS & PESTS

Occasional problems with powdery mildew, stem canker, rust, leaf spot, leafminers and caterpillars are possible. Slugs will eat the buds off of Creeping Phlox.

The name phlox is derived from the Greek word meaning 'a flame.'

P. paniculata

Pinks
Dianthus spp.

Flower color: Pink, red, white or lilac purple.
Height: 6–16" (15–40 cm) tall. **Spread:** 12–24" (30–60 cm) wide.
Blooms: Spring and summer.

These beautiful little plants are valued as much for their spicy fragrance as for their delicate flowers. Pinks aren't named because of their color. The ruffled, uneven edges of the petals look as though they have been trimmed with pinking shears.

PLANTING

Seeding: Easy to grow from seeds started in cold frame in fall or early spring.

Planting out: Spring.

Spacing: 10–20" (25–50 cm) apart.

GROWING

A location with **full sun** is preferable, but some light shade will be tolerated. A **neutral or alkaline** soil is required. The most important factor in the successful cultivation of pinks is drainage—they hate to stand in water. Mix sand or gravel into their area of the flowerbed to encourage good drainage. Pinks make excellent plants for rock gardens, rock walls and for edging flower borders and walkways. Frequent division, each year or two, will keep the plants vigorous. Divide plants in early spring.

RECOMMENDED

D. deltoides (Maiden Pinks) grow from 6–12" (15–30 cm) tall and about 12" (30 cm) wide. The plant forms a mat of foliage and will flower in spring. This is a popular species in rock gardens.

D. gratianopolitamus (Cheddar Pinks) usually grow about 6" (15 cm) tall, but can grow up to 12" (30 cm) tall and 18–24" (45–60 cm) wide. This plant is

The tiny, delicate petals of pinks can be used to decorate cakes. Be sure to remove the white part at the base of the petal before using the petals or they will be bitter.

long-lived and forms a very dense mat of evergreen, silvery gray foliage with sweet-scented flowers borne in summer.

D. plumarius (Cottage Pinks) is noteworthy for its role in the development of many popular cultivars known as Garden Pinks. They are generally 12–18" (30–45 cm) tall and 24" (60 cm) wide, although smaller cultivars are available. They all flower in spring and into summer if deadheaded regularly. The flowers can be single, semi-double or fully double and are available in many colors.

D. plumarius

GARDENING TIPS

Pinks may be difficult to propagate by division. It is often easier to take cuttings in summer, once flowering has finished. Cuttings should be 1–3" (2.5–8 cm) long. Strip the lower leaves from the cutting. The cuttings should be kept humid, but be sure to give them some ventilation so that fungus problems do not set in.

Deadheading the plants as the flowers fade is a good way to prolong the blooming period. Leave a few flowers in place to go to seed towards the end of the blooming period. The plants will self-seed quite easily. Seedlings may differ from the parent plants, often with new and interesting results.

PROBLEMS & PESTS

Providing good drainage and air circulation will keep most of the fungus problems away. Occasional problems with slugs, blister beetles, sow bugs, grasshoppers, chipmunks, squirrels and deer are possible.

The Cheddar Pink is a rare and protected species in Britain. It was discovered in the 18th century by Wiltshire botanist, Samuel Brewer, and it became as locally famous as Cheddar cheese.

D. deltoides

Primrose

Primula spp.

Flower color: Red, orange, pink, purple, blue, white or yellow.
Height: 6–24" (15–60 cm) tall. **Spread:** 6–18" (15–45 cm) wide.
Blooms: Spring and early summer.

Primroses welcome spring with their soft and velvety looking blooms, and the 'primrose path' is proverbially seductive. Don't stop with just primroses alongside your garden path however. Partner primroses with dwarf daffodils, lungworts and pulmonarias for a tapestry of woodland flowers that are the jewels of the spring display.

PLANTING

Seeding: Direct sow in garden at any time of year. Start indoors in early spring or in cold frame in fall or late winter.

Planting out: Spring.

Spacing: 6–18" (15–45 cm) apart.

GROWING

Choose a location for these plants with **partial shade**. Soil should be **moderately fertile, humus-rich, moist, well drained and neutral or slightly acidic**. Primroses may be incorporated into many areas of the garden. There are moisture-loving primroses that may be included in a bog garden or grown in a moist location. Woodland Primrose grows well in a woodland garden or under the shade of taller shrubs and perennials in a border or rock garden. The species with flowers on tall stems look excellent when planted in masses, while the species with solitary flowers peeking out from amongst the foliage are interesting when dotted throughout the garden in odd spots. Overgrown clumps should be divided after flowering or in early fall.

Polyantha hybrid

This herald of spring gets its name from the Latin words prima *meaning 'first' and* rosa *meaning 'rose.'*

P. eliator

RECOMMENDED

P. eliator (Oxlip) grows about 12" (30 cm) tall and 6" (15 cm) wide. The yellow tubular flowers are clustered at the ends of long stems.

P. japonica (Japanese Primrose) grow from 12–24" (30–60 cm) tall and from 12–18" (30–45 cm) wide. Japanese Primrose will thrive in moist, boggy conditions. It is a candelabra flowering type, which means that the long flower stem will have up to six evenly spaced rings of flowers along its length.

P. vulgaris (Common Primrose; English Primrose) grows 6–8" (15–20 cm) tall and 8" (20 cm) wide. The flowers are solitary and borne at the ends of short stems that are slightly longer than the leaves.

There are several hybrids that have been developed by crossing different species of primrose. The most popular are the **Polyantha hybrids**. Available in a wide range of solid or bi-colors, Polyantha hybrids usually grow 8–12" (20–30 cm) tall with about an equal spread. The flowers are clustered at the tops of stems of variable height.

GARDENING TIPS

Pull off yellowing or dried leaves in fall for fresh new growth in spring.

Primroses, like many woodland plants, have a preferred temperature zone. If the temperature is too hot or too cold they will stop blooming. Plant them in a variety of sites to increase your chances of having content flowers.

PROBLEMS & PESTS

Slugs and snails are likely to cause the biggest problems. Other possible problems may be caused by strawberry root weevils, aphids, rust or leaf spot.

Concoctions were made from primrose in the Middle Ages and used as love potions and, coincidentally, to relieve headaches.

P. japonica

Purple Coneflower

Echinacea purpurea

Flower color: Purple, pink, mauve or white. Rusty orange centers.
Height: About 60" (150 cm) tall. **Spread:** 18" (45 cm) wide.
Blooms: Summer.

A great perennial for the beginning gardener, Purple Coneflower has had known health benefits for generations but can be considered particularly healthy for the soul of the gardener because it is dependable and beautiful, often blooming during the hottest days of summer when other flowers have lost their flush of freshness. The large daisy-like blooms look right at home displayed in front of a rustic split-rail fence or alongside the weather-beaten, sunny wall of a barn or garden shed.

PLANTING

Seeding: Direct sow in garden in spring.

Planting out: Spring.

Spacing: 18" (45 cm) apart.

GROWING

Coneflower needs to be in a location that receives **full sun.** Soil should be **average to fertile,** but poor soil is tolerated with only a slight decrease in vigor. Good drainage is essential. Coneflower prefers to have regular water but is drought-resistant because of its thick taproot for storing water. This North American native looks good in meadow gardens, in light, open woodlands or in an informal border. Divide every four years in spring or fall.

RECOMMENDED

'Magnus' is 36" (90 cm) tall. Its large flowers are up to 7" (18 cm) across, in deep purple, with dark orange centers.

'White Lustre' has cream-colored flowers with brownish-orange centers.

GARDENING TIPS

Deadheading early in the season is recommended as it prolongs the flowering season. Later in the season you may wish to leave the flowerheads in place. The dry flowerheads make an interesting feature in fall and winter gardens. The plants may self-seed, providing more plants. If you don't want them to self-seed then remove all the flowerheads as they fade.

PROBLEMS & PESTS

Powdery mildew is the most common problem, but Purple Coneflower may also get leafminers, bacterial spots or gray mold. The roots are sometimes attacked by vine weevils.

Echinacea was discovered by the First Nations Peoples and was one of their most important medicines. It is a popular immunity booster in herbal medicine today.

'Magnus'

'White Lustre'

Red-Hot Poker
Common Torch Lily
Kniphofia uvaria

Flower color: Bright red to orange to pale yellow or cream.
Height: About 48" (120 cm) tall. Different cultivars may be taller or shorter than this.
Spread: 24" (60 cm) wide. Some variation in the cultivars.
Blooms: Early summer to fall.

If you are looking for something different, something exotic, something to really heat up the garden beds or make a hot fashion statement, then don't be cool to the demands of Red-hot Poker. The intense, fiery colors will brand a memory into your mind and sear the garden scene with startling exclamation points. Red-hot Poker is a bit more temperamental about winter protection than other perennials, but, for the extroverted gardener, it is worth the effort.

PLANTING

Seeding: Straight species may be started in cold frame in early spring. Cultivars are unlikely to come true to type.

Planting out: Spring.

Spacing: 24" (60 cm) apart.

GROWING

These plants will grow equally well in **full sun or partial shade**. Soil should be **fertile, humus-rich, sandy and moist**. Red-hot Poker makes a bold, vertical statement in the middle or back of a border. These plants look best when grouped together. Large clumps may be divided in late spring. Do not cut back the foliage in fall because it is needed for winter protection for the plants.

RECOMMENDED

'Little Maid' grows to only 24" (60 cm) tall and has salmon-colored buds opening to white flowers.

'Nobilis' is a large plant growing up to 72" (180 cm) tall with long orange-red flower spikes, borne from midsummer to fall.

GARDENING TIPS

To encourage the plants to continue flowering for as long as possible, cut the spent flowers off right where the stem meets the plant.

The plants are sensitive to cold, wet weather. Bundle up the leaves and tie them above the crown in fall to keep it dry and more protected.

PROBLEMS & PESTS

This plant rarely has any problems but is susceptible to stem or crown rot. Flowers dropping off unopened may be caused by thrips.

Often there will seem to be two colors on one spike because the buds and matured flowers are different colors.

Rock Cress
Common Aubrieta
Aubrieta deltoidea

Flower color: White, rose pink or purple.
Height: 2–6" (5–15 cm) tall. **Spread:** 24" (60 cm) or wider.
Blooms: Spring or early summer.

It took me years of research and testing, but I finally discovered a rock-garden plant that will bloom in my partially shaded rock outcropping and survive the onslaught of slugs. Rock Cress is the wonder plant that did the job with the fuzzy gray leaves that are less attractive to slugs than other groundhugging bloomers and less demanding of constant sunshine. The richly colored blooms are as soft looking as the foliage, and the growth habit is tidy and low. It looks great covering larger boulders.

PLANTING

Seeding: Start seeds in cold frame in spring or fall. May not come true to type.

Planting out: Fall.

Spacing: 18" (45 cm) apart.

GROWING

Rock Cress prefers **full sun**, but will tolerate partial shade. Soil should be of **average fertility, well drained with rocks or gravel** mixed in. It also prefers soil to be a little on the alkaline side. Use Rock Cress in the crevices of a rock wall, between the paving stones of a pathway, in a rock garden, along the edge of a border or beneath taller plants. Every year or two, in fall, Rock Cress will need dividing in order to prevent the clump from thinning and dying out in the middle. Shear the old flowers because seeds will not bloom true to form.

This low-growing spreader can cover large areas but rarely becomes invasive.

RECOMMENDED

'**Purple Cascade**' has purple flowers.

'**Red Carpet**' has rosy red flowers.

'**Variegata**' has purple-blue flowers and gold and green variegated leaves.

'**Whitewell Gem**' has purple blooms.

GARDENING TIPS

Rock Cress should be sheared back by half once it has finished flowering. This will encourage compact growth and may occasionally result in a second flush of flowers later in the season.

PROBLEMS & PESTS

Sometimes Rock Cress has trouble with aphids, nematodes and flea beetles.

Rock Cress is very popular in England where its purple cascades of flowers slip over rock walls and brighten up the rainiest spring day.

Rose Campion

Lychnis coronaria

Flower color: Magenta, purple or white
Height: 24–36" (60–90 cm) tall. **Spread:** 18" (45 cm) wide.
Blooms: Early to late summer.

The hot pink blooms of this gray-leaved perennial are small but so intense in color that they were once banished from the gardens of color-conscious designers. Garden writers a generation ago insisted that magenta was a color that clashed severely with other flower shades. Thank goodness those days of color snobbery are over and anyone can gleefully add the tropical punch of this easy, self-seeding perennial to the garden. Watch it explode with saturated color next to bright orange daylilies and lime green *Euphorbia* like the festival chaos on a Hawaiian-print shirt. If loud and brash colors are not your style, grow *Lychnis* with soft pinks and lavender purples to tone it down a bit or use the subdued and less vigorous white and pastel varieties.

PLANTING

Seeding: Start seeds in fall or late spring.

Planting out: Spring or fall.

Spacing: 18" (45 cm) apart.

GROWING

Rose Campion will grow equally well in **full sun or partial shade**. Soil should be of **average fertility and well drained**. Use Rose Campion in an informal or cottage garden or in a wildflower garden.

RECOMMENDED

'**Alba**' has white flowers.

'**Angel's Blush**' has white flowers with cherry red centers.

ALTERNATE SPECIES

L. chalcedonica (Maltese Cross) grows 36–48" (90–120 cm) tall and about 12–18" (30–45 cm) wide. The scarlet flowers are in clusters. This species definitely needs some support.

L. chalcedonica

GARDENING TIPS

These tall plants may need some support, particularly if they are in a windy location. Peony supports or twiggy branches pushed into the soil before the plants get too tall work best and are less noticeable than having the plants tied to stakes.

Rose Campion is quite short-lived. It does tend to self-seed though and will repopulate the garden with new plants as old ones die out. It reseeds readily in gravel pathways where young plants are easily transplanted.

These bright and lively flowers may turn up, through self-seeding, where you least expect them.

Children love to string together these bright pink flowers like daisy chains.

Rose-Mallow
Hibiscus moscheutos

Flower color: White, red or pink.
Height: About 96" (245 cm) tall. **Spread:** 36" (90 cm) wide.
Blooms: Late summer to frost.

Add a touch of the tropics to your garden with the huge blooms and exotic look of the Rose-mallow. I use this tall, shrubby perennial at the base of a rocky slope where the drainage water soaks the roots and the wall behind provides wind protection for the flower-laden stalks. The late-summer bloom time is a welcome refreshment as other perennials start to fade.

PLANTING

Seeding: Sow seeds in spring. Ensure soil temperature is 55–64° F (13–18° C).

Planting out: Spring.

Spacing: 36" (90 cm) apart.

GROWING

Grow Rose-mallow in **full sun**. Soil should be **humus-rich, moist and well drained**. This is an interesting plant for the back of an informal border or mixed into a pond planting. The large flowers, often the size of dinner plates, make very colorful additions to late summer gardens. They create a bold focal point when in flower. Deadhead spent flowers to keep the plant tidy. Divide in spring.

RECOMMENDED

There are several compact varieties available. They grow to about half the height of the straight species, but still have huge 8–10" (20–25 cm) flowers.

'Blue River II' has blue-tinged leaves and white flowers.

'Lady Baltimore' has bright pink flowers with dark red eyes.

'Lord Baltimore' has bright red flowers with ruffled petals.

'Southern Belle' has deep pink, red, light pink or white flowers.

'Blue River II'

GARDENING TIPS

Prune plants by one-half in June for bushier, more compact plants.

Rose-mallow grows best where summers are hot and seasons are clearly defined. Mulch the first winter.

PROBLEMS & PESTS

Rose-mallow may develop problems with rust, fungal leaf spots, bacterial blight, *Verticillium* wilt, viruses, and stem or root rot. A few possible insect pests are whiteflies, aphids, scale insects, mites and caterpillars.

The moisture-loving Rose-mallow is one of the most exotic-looking plants you can include in a bog-garden or pondside planting.

'Southern Belle'

Russian Sage

Perovskia atriplicifolia

Flower color: Blue or purple.
Height: 48" (120 cm) tall. **Spread:** 36" (90 cm) wide.
Blooms: Midsummer to fall.

No need to improve your poor, rocky soil if you invite Russian Sage to provide its vertical accent and gray-leaved contrast, although this plant will grow even better in soil that is loosened and improved a bit. Use this pale beauty in an all white or pastel garden or with other herbs that have interesting foliage and the fragrant, blue blooms will be an added attraction in late summer.

PLANTING

Seeding: Not recommended.

Planting out: Spring.

Spacing: 36" (90 cm) apart.

The airy habit of this plant creates a mist of silver purple in the garden.

GROWING

Russian Sage prefers a location with **full sun**. Soil should be **poor to moderately fertile and well drained**. This plant has the potential to rot in wet winters. The silvery foliage and blue flowers combine well with other plants in the back of a mixed border that softens the appearance of Daylily. Russian Sage may also be used in a natural garden or on a dry bank. Russian Sage does not need dividing.

RECOMMENDED

'Blue Spire' is an upright plant with deep blue flowers and feathery leaves.

'Filagran' has delicate foliage and an upright habit.

'Longin' is narrow and erect and has more smoothly edged leaves than other species.

GARDENING TIPS

Cut the plant back to 12" (30 cm) in fall or winter to keep summer growth fresh and vigorous.

'Filagran'

Sandwort

Arenaria montana

Flower color: White flowers with yellow eyes.
Height: 1–4" (2.5–10 cm) tall. **Spread:** 12" (30 cm) wide.
Blooms: Late spring or early summer.

Remember the 'sand' in the name and add sand to the planting hole. This plant can be put to work doing your weeding; I use all the varieties between stepping stones and bricks that were set on a sandy base. Its ground-hugging nature smothers weeds and then pops up in unexpected cracks and crevices to guard against any invasive ideas that weed seeds may have. The plants are easy to pull up if they get a little too enthusiastic.

PLANTING

Seeding: Plant seeds outdoors in fall.

Planting out: Spring or fall.

Spacing: About 10" (25 cm) apart.

GROWING

Sandwort likes to grow in **full sun or light shade**. Soil should be **poor or average, sandy and well drained**. Sandwort does well in a rock garden, stone wall or between the paving stones of a path. Divide in spring or fall, whenever the center of the plant begins to thin out.

ALTERNATE SPECIES

A. balearica (Corsican Sandwort) is a 3" (8 cm) tall, dense mat. It requires moist soil and has small white blooms.

A. verna (Moss Sandwort) has evergreen, moss-like foliage. It works well between stepping stones. *A.v.* **var.** *caespitosa* (Irish Moss) is 2" (5 cm) tall with star-shaped, white flowers.

GARDENING TIPS

Sandwort likes to be watered regularly. To avoid spending all summer using the hose or watering can it is best to plant Sandwort beneath taller shrubs and to apply a mulch to the soil surface. Sandwort has shallow roots and will not compete with the larger shrubs. Both the mulch and the Sandwort will protect the roots of the bigger shrub.

Sandwort requires cold temperatures in winter. The best way to get it to flourish is to plant it in the coldest part of the garden. The bottom of a hill or a cold dip where water drains freely, is an ideal location.

PROBLEMS & PESTS

Occasional problems with rust or anther smut are possible.

Sandwort forms an attractive mat of foliage that is inviting to bare feet.

A. montana

A.v. var. caespitosa

Saxifrage
Saxifraga spp.

Flower color: Red, white, yellow or pink.
Height: About 6–24" (15–60 cm) tall. **Spread:** 6–24" (15–60 cm) wide.
Blooms: Varies from early spring to early summer.

The Latin meaning for this plant's name is 'breaks rocks' and if you wish to fill a loose rock wall with something that will crowd out grassy weeds and other invaders, add the mossy saxifrage or the species London Pride for carefree spring color. The tall blooming spikes of London Pride are long lasting yet delicate—perfect for adding to a bud vase where it will outlast several rose buds before it drops its petals. One of the most under-used but easy to grow perennials for the shady garden.

PLANTING

Seeding: Sow fresh seeds in cold frame.

Planting out: Spring.

Spacing: 6–24" (15–60 cm) apart.

GROWING

Saxifrage prefers to be planted in **partial shade.** Soil should be **neutral to alkaline, fertile, moist and well drained.** These plants make excellent additions to rock gardens or to edge beds and borders. Divide in spring.

RECOMMENDED

S. paniculata forms a 6" (15 cm) tall and 12" (30 cm) wide cushion of leaves. The edges of the leaves are usually lime-encrusted. In early summer long spikes of clustered flowers extend above the plant, usually in white. This species enjoys more sun.

S. stolonifera (Strawberry Begonia) is used as a groundcover in moist soil. The leaves are attractively veined and the tiny white flowers are borne on spikes. The parent plant sends out shoots at the end of which grow tiny new plants. This plant is also popular for hanging baskets.

S. x urbium (London Pride) is one of the easiest saxifrages to grow. It grows about 12" (30 cm) tall and 18" (45 cm) wide. This species makes an excellent groundcover and bed edger. It tolerates the pollution of urban conditions. The flowers are usually white, though there are cultivars in shades of pink.

GARDENING TIPS

The small rosettes of leaves that form around the base of the plant may be removed and rooted in late spring or early summer. *S. stolonifera* is popular with children because the young plants that grow at the ends of the stolons can be detached and will root quickly in pots.

PROBLEMS & PESTS

Saxifrage may have occasional trouble with aphids, slugs, vine weevil grubs or spider mites.

There are over 400 species of Saxifrage and even more cultivars. A complicated system was developed to classify the many forms by grouping the plants into different sections, subsections and series.

S. x urbium

Scabiosa
Pincushion Flower
Scabiosa caucasica

Flower color: Lavender blue, also white, pink or deep blue.
Height: 24" (60 cm) tall. **Spread:** 24" (60 cm) wide.
Blooms: Summer.

Both the rounded, dome-shaped flower buds and the spent seedheads look like pin cushions, but it is the repeat blooming and sky blue and lavender purple shades that make this such an endearing perennial for butterfly enthusiasts and cottage gardeners. I grow this spring and summer bloomer in a raised rockery bed where it won't rot from winter rains, and it rewards me with flowers from May until October, as long as I remember to continue cutting the blooms.

PLANTING

Seeding: Start freshly ripened seeds in cold frame in fall or by spring.

Planting out: Spring.

Spacing: 24" (60 cm) apart.

GROWING

This plant prefers **full sun** but will tolerate partial shade. Soil should be **light, moderately fertile, neutral or alkaline and well drained**. These plants look best when they are planted in groups in a bed or border. They are also used as cut flowers. Divide in early spring whenever the clumps become overgrown.

RECOMMENDED

'Butterfly Blue' is 18" (45 cm) tall with lavender blue flowers.

'Fama' has sky blue flowers with silvery white centers.

'Miss Wilmott' has white flowers.

'Pink Mist' is 12" (30 cm) tall and is more compact with more blooms than other cultivars.

GARDENING TIPS

Remove the flowers as they fade to promote a longer flowering period. Cutting flowers at their peak every few days for indoor use will make this maintenance chore more enjoyable.

PROBLEMS & PESTS

Scabiosa rarely has any problems. Sometimes aphids can be troublesome.

'Miss Wilmott'

The genus name, Scabiosa, *is from the historical use of this plant to treat scabies.*

Sea Holly
Eryngium spp.

Flower color: Purple, blue or white.
Height: 12–48" (30–120 cm) tall. **Spread:** 12–24" (30–60 cm) wide.
Blooms: Summer and fall.

Slate blue foliage and prickly leaves make this a distinctive plant that can handle poor, dry soil, and pairing it with other drought-tolerant foliage plants such as artemisias, succulents and Russian Sage will create a garden area with fabulous texture that can survive neglect even in August when summer vacations make watering the garden a difficult task.

PLANTING

Seeding: Direct sow in garden.

Planting out: Spring.

Spacing: 12–24" (30–60 cm) apart.

GROWING

Grow sea holly in a location with **full sun**. Soil should be **average to fertile and well drained**. The long tap root makes sea holly fairly drought-tolerant but adverse to prolonged periods without water. The long-lasting flowers look attractive with other flowers in a border. Sea holly also makes an interesting addition to naturalized gardens. This plant is very slow to re-establish itself after dividing. Root cuttings should be taken in late winter.

E. giganteum

The roots of one species of sea holly were candied and used as an aphrodisiac lozenge known as an Eryngoe. Don't expect to find them today as they were only made several centuries ago in England.

RECOMMENDED

E. alpinum (Alpine Sea Holly) grows 24–48" (60–120 cm) tall. This species has soft, spiny bracts and steel blue or white flowers. There are several cultivars available in different shades of blue.

E. giganteum (Giant Sea Holly) grows 48–60" (120–150 cm) tall. The flowers are steel blue with silvery gray bracts.

E. x tripartitum grows 24–36" (60–90 cm) tall. The flowers are purple and the bracts are gray tinged with purple.

E. varifolium (Moroccan Sea Holly) grows 12–16" (30–40 cm) tall. It has dark green leaves with silvery veins and gray-purple flowers with blue bracts.

E. alpinum

GARDENING TIPS

The long-lasting flowers are distinctive when used in fresh or dried arrangements. Remove the flower-heads from their short stems and attach them to longer florist's wire for a successful arrangement.

PROBLEMS & PESTS

Roots may rot if the plants are left in standing water for prolonged periods of time. Slugs, snails and powdery mildew may be problems.

Some species of sea holly are used as flavoring for jams and toffees and in Latin American dishes.

E. × tripartitum

Sea Pink
Common Thrift
Armeria maritima

Flower color: Pink or white.
Height: About 8" (20 cm) tall. **Spread:** About 12" (30 cm) wide.
Blooms: Late spring or early summer.

A plant with an international flavor, Sea Pink is found in mountains and along the coastlines across most of the Northern Hemisphere. It is the perfect plant for growing in pots at an oceanside cabin where it can survive neglect or for adding tufts of texture to a sunny rock garden or alongside a flagstone path. The rounded pink blossoms that are held on erect stems blend well with other drought-tolerant plants with gray foliage.

PLANTING

Seeding: Start seeds in cold frame in spring or fall.

Planting out: Spring.

Spacing: 10" (25 cm) apart.

GROWING

Sea Pink requires **full sun.** Soil should be **poor or moderate and well drained.** Sea Pink is very drought-tolerant. This is a great plant for seaside gardens because it is tolerant of salt spray. It may also be used in rock gardens or in the front of a border. Divide in spring or fall.

RECOMMENDED

'Alba' has flowers in white.

'Bloodstone' has flowers so dark they appear to be blood red.

'Dusseldorf Pride' has rose pink flowers.

'Vindictive' is a smaller plant growing to only 6" (15 cm) and with purple-pink flowers.

GARDENING TIPS

If your Sea Pink seems to be dying out in the middle of the clump, then try cutting it back hard. New shoots should fill in quickly.

PROBLEMS & PESTS

Problems are rare with this durable plant. It may occasionally get rust or be attacked by aphids.

Attract bees and butterflies to the seaside garden with clumps of Sea Pink.

Sedum

Stonecrop
Sedum spp.

Flower color: Yellow, white, red or pink.
Height: 4–24" (10–60 cm) tall. **Spread:** 12–18" (30–45 cm) wide.
Blooms: Summer, late summer or fall.

These plants come in many different forms. Some form upright clumps of woody stems while others form low creeping mats. All generally have fleshy, succulent foliage. The sedums are real problem solvers for the difficult growing conditions found in the dry, sandy soil favored by our native cedar trees.

PLANTING

Seeding: Start seeds in cold frame in fall.

Planting out: Spring.

Spacing: About 18" (45 cm) apart.

GROWING

These plants prefer **full sun** but will tolerate partial shade. Soil should be **average, very well drained and neutral to alkaline**. Low-growing sedums make excellent groundcovers and rock-garden or rock-wall plants. They also edge beds and borders wonderfully. The taller types give a beautiful late season display in a bed or border. Divide in spring, when needed.

RECOMMENDED

'Autumn Joy' (Autumn Joy Sedum) is a popular, upright hybrid. The flowers open in pink or red and later fade to deep bronze. The plant forms a clump 24" (60 cm) tall and wide.

S. acre (Gold Moss Stonecrop) grows 2" (5 cm) tall and spreads indefinitely. The small yellow-green flowers are borne in summer.

S. rupestre

S. spurium

Low-growing sedums make an excellent groundcover under trees. Their shallow roots survive well in the competition for space and moisture.

S. rupestre (Stone Orpine) is a mat-forming species that grows to about 4" (10 cm) tall and spreads 24" (60 cm) wide.

S. spathifolium is a mat-forming species with pale blue-green leaves. It grows to about 4" (10 cm) tall and spreads 24" (60 cm) wide.

S. spectabile (Showy Stonecrop) is an upright species with pink flowers borne in late summer. It forms a clump 18" (45 cm) tall and wide.

S. spurium (Two-row Stonecrop) forms a mat about 4" (10 cm) tall and 24" (60 cm) wide. The flowers are deep pink or white.

S. spectabile

S. spathifolium

GARDENING TIPS

Prune back 'Autumn Joy' in May by one-half and insert pruned-off parts into soft soil. Cuttings will root quickly and early summer pruning makes for compact, bushy plants.

PROBLEMS & PESTS

Slugs, snails and scale insects may cause trouble for these plants.

'Autumn Joy'

Autumn Joy Sedum brings color to the late season garden, when few flowers are in bloom.

Snow-in-Summer
Cerastium tomentosum

Flower color: White.
Height: 4–8" (10–20 cm) tall. **Spread:** Indefinite.
Blooms: Late spring or early summer.

When in bloom, in late spring and early summer, Snow-in-summer is completely covered by small, white, five-petalled flowers. The foliage is silvery white and covered in little hairs. It contrasts dramatically with plants bearing green foliage and brightly colored flowers. This plant may be quite invasive, spreading rapidly outwards in even the most difficult conditions. This quality may be a benefit or a detriment depending on what you want the plant to do.

PLANTING

Seeding: Start seeds early indoors or direct sow in garden for flowers in first year.

Planting out: Spring.

Spacing: 12–18" (30–45 cm) apart.

GROWING

Grow Snow-in-summer in **full sun or partial shade**. This plant will grow in any type of **well-drained** soil but may develop root-rot in wet soil. The richer the soil, the more invasive Snow-in-Summer becomes, but will do well in poor soil.

Snow-in-summer is well suited to sunny, hot, well-drained locations. It may be used under taller plants, as a ground cover, along border edges and to prevent erosion on sloping banks. It is attractive in a rock wall, but its invasiveness may overwhelm the often smaller and less vigorous plants that are popular in rock walls.

This bright, white plant has a dark side. Its vigorous nature can quickly move into the invasive category. If left to grow unchecked, Snow-in-summer can take over entire flowerbeds. Some gardeners have banished it completely from the garden after it successfully takes over the territory of less vigorous plants. Snow-in-summer tends to die out in the middle as it grows, so dividing it every two years will ensure that it maintains even coverage where you want it to.

RECOMMENDED

'Silver Carpet' is a more compact cultivar.

C. t. var. *columnae* is a shrubbier, less spreading variety.

Snow-in-summer is a good choice on dry slopes or in beds under eaves, in areas where little else thrives. Planting it in areas such as this will help to overcome its tendency to become invasive in the more lush areas of the garden.

GARDENING TIPS

Cutting the plant back after it has finished flowering and again later in summer will help keep growth in check and prevent the plant from thinning out excessively in the center. It can grow up to 36" (90 cm) in a single year. The plant roots in two ways: the roots send up new shoots as they grow and the sprawling stems send down new roots. This makes taking cuttings easy: just cut off stem ends that have already started rooting and plant them where you want them.

PROBLEMS & PESTS

Slugs may nibble off the flower buds.

The silvery foliage and abundant white flowers contrast nicely with the brilliant blue flowers of Forget-me-not.

Spike Speedwell
Veronica
Veronica spicata

Flower color: White, pink, purple or blue.
Height: 12–24" (30–60 cm) tall. **Spread:** 16–18" (40–45 cm) wide.
Blooms: Summer.

Long-lasting blooms on fuzzy spikes of white, pink or purple let Spike Speedwell fill in the empty spots of a lightly shaded woodland garden for carefree summer color.

The flowers of this plant fade quickly once picked giving this plant its common name, speedwell, which was a word used to wish people a safe and quick journey.

PLANTING

Seeding: Start seeds in cold frame in fall.

Planting out: Spring.

Spacing: About 18" (45 cm) apart.

GROWING

Speedwells prefer **full sun**, but they will tolerate partial shade. Soil should be of **average fertility, moist and well drained**. Plant *V. spicata* in masses in a bed or border. An alternative species, Prostrate Speedwell, is useful in a rock garden or at the front of a border. Divide speedwells in fall or spring every three or five years.

RECOMMENDED

There are many available cultivars and varieties.

'Icicle' has white flowers.

'Red Fox' has dark red-pink flowers.

V. s. **ssp.** *incana* has soft, hairy, silvery green leaves and deep purple-blue flowers.

ALTERNATE SPECIES

An interesting hybrid is *V.* **x 'Sunny Border Blue'** with rich green, crinkled foliage and sturdy spikes of royal blue.

V. prostrata (Prostrate Speedwell) is 6" (15 cm) tall. **'London Blue'** has bright blue flowers and grows up to 8" (20 cm) tall. **'Mrs. Holt'** has light pink flowers. **'Trehane'** has yellow or golden-hued leaves and deep blue flowers. Use Prostrate Speedwell in a rock garden or along the front of a border.

GARDENING TIPS

Deadhead to encourage a longer blooming period. For tidier plants, shear back the tall types to about 6" (15 cm) in June.

PROBLEMS & PESTS

Problems with scale insects are possible, as are fungus problems like downy mildew, powdery mildew, rust, leaf smut and root rot.

V.s. ssp. *incana*

'Red Fox'

Stonecress
Aethionema spp.

Flower color: Red, pink, cream or white.
Height: 2–12" (5–30 cm) tall. **Spread:** 4–12" (10–30 cm) wide.
Blooms: Spring to early summer.

These small shrubby perennials have gray-green or blue-green leaves that look attractive all summer. In late spring or early summer the plants are covered by clusters of small, star-shaped flowers. Stonecress like the sun and don't need much water, but they are somewhat adverse to hot weather.

PLANTING

Seeding: Direct sow in garden in spring, or start early in cold frame.

Planting out: Early spring.

Spacing: 3–10" (8–25 cm) apart.

GROWING

Stonecress must be planted in a location that receives **full sun**. This plant will tolerate any type of soil, but prefers **well-drained, alkaline** soil. It requires only moderate watering and will tolerate some drought, if the weather is not too hot. It will do best if not exposed to excesses of heat, preferring a slightly cool summer climate. Stonecress is a fairly low-growing plant that spreads only modestly. It makes a wonderful addition to stone walls and rock gardens. These perennials are quite short-lived. Division is unlikely to be required.

RECOMMENDED

A. armenum grows 6–8" (15–20 cm) tall and wide. It bears small clusters of pale pink flowers.

A. grandiflorum grows 8–12" (20–30 cm) tall, with an equal spread. It bears pale to deep rose pink flowers.

A. oppositifolium grows 2" (5 cm) tall and 4–6" (10–15 cm) wide. It bears lavender pink flowers. A low-growing, mat forming species of stonecress.

A. schistosum grows 5–10" (13–25 cm) tall, with an equal spread. It bears fragrant, rose-colored flowers.

A. warleyense (Warley Rose Stonecress) grows 6–8" (15–20 cm) tall, with an equal spread. It bears pink flowers. This is the most heat-tolerant species of them all.

PROBLEMS & PESTS

Aphids and spider mites may be problems.

Plants are likely to self-seed. Deadheading is helpful in preventing dozens of plants from cropping up all over the flowerbed.

A. warleyense

GARDENING TIPS

Stonecress is a short-lived perennial, but, because it grows readily from seed, it is possible to economically replace dying plants with new ones. Seeds collected from garden plants may not come true to type. Taking cuttings from new growth, before flowering begins or after it has finished, is another good way to keep a supply of new plants to replace the old.

Deadhead the plant once it has finished flowering in order to keep the plant looking compact and neat. Take a sharp pair of garden shears and trim the plant, removing the spent flowers as you go. Plants that are not deadheaded will become leggy and lose their compact appearance.

If the plant is suffering from too much heat, try mulching the soil around the plant to keep the roots cool.

Sweet Rocket
Dame's Rocket
Hesperis matronalis

Flower color: Magenta, pink, white or purple.
Height: About 36" (90 cm) tall. **Spread:** 18" (45 cm) wide.
Blooms: Spring.

Tall, fragrant blooms that seem to love cool, mild weather make this one perennial that won't sulk when Northwest summers stay cloudy and wet. The rocket-like blooming spires come in a deep purple that combines well with the bright yellow blooms of daylily and looks lovely supported with a foreground planting of peony. Sweet Rocket's flowers, and their scent, last quite well when used in fresh flower arrangements.

PLANTING

Seeding: Direct sow in garden in spring. Flowers will bloom in following year.

Planting out: Spring or fall.

Spacing: 18" (45 cm) apart.

GROWING

Grow these plants in **full sun or partial shade.** Soil should be **average to fertile, humus-rich, moist and well drained.** These are excellent plants to include in a woodland garden or meadow planting. They look good in a mixed border with other spring flowering plants. Be sure there are late-flowering plants mixed in nearby as *Hesperis* may go dormant by midsummer. These short-lived perennials are unlikely to need dividing. They self-seed easily and will provide a constant supply of new plants.

RECOMMENDED

'**Purpurea Plena**' has purple, double flowers.

H. m. var. *albiflora* has white flowers. '**Alba Plena**' has white, double flowers.

GARDENING TIPS

Some population control may be needed once these plants get going. Thin out seedlings and trim back some of the plants before they set seed if too many offspring are beginning to show up in the garden. Double-flowered varieties will not set seed but may be propagated from cuttings taken from near the base of the plant in spring.

PROBLEMS & PESTS

Occasional problems with viruses, mildew, slugs, snails, flea beetles and caterpillars are possible.

This cottage garden favorite adds its sweet scent to a pleasant summer evening in your garden. Dried flowers of Sweet Rocket also make a good addition to potpourri.

Often found growing wild in much of the temperate world as a garden escapee Sweet Rocket is actually indigenous to Italy.

Sweet Woodruff

Galium odoratum

Flower color: White.
Height: About 12" (30 cm) tall. **Spread:** May spread indefinitely.
Blooms: Late spring to midsummer.

I use Sweet Woodruff as a lacy petticoat poking out beneath woodland ferns and among the shade-loving primroses. Very easy to grow, they provide early color before the Astilbes and hydrangeas fill the shaded areas with summer color. The light and airy flowers have a delicate Baby's Breath quality that never fails to add a fresh look to the garden.

PLANTING

Seeding: Start in cold frame in late summer or fall.

Planting out: Spring or fall.

Spacing: About 12" (30 cm) apart.

GROWING

This plant will grow well in full shade but prefers **partial shade**. Soil should be **humus-rich and evenly moist**. Sweet Woodruff is a perfect woodland groundcover. It loves the same conditions in which azaleas and rhododendrons thrive and forms a beautiful green carpet. Divide plants in spring or fall.

GARDENING TIPS

Shear back after blooming to encourage plants to fill in with foliage and crowd out weeds. Sweet Woodruff will not bloom well in full shade. This plant can become invasive.

PROBLEMS & PESTS

May have problems with downy mildew, powdery mildew, rust and fungal leaf spots.

The dried leaves were once used to scent doorways and freshen stale rooms.

The wiry, creeping habit of Sweet Woodruff makes it an excellent groundcover; it is particularly effective under trees. The white flowers seem to glow at dusk in late spring.

Thyme

Thymus spp.

Flower color: Purple, pink or white.
Height: 6–12" (15–30 cm) tall. **Spread:** 16" (40 cm) wide.
Blooms: Late spring and early summer.

This popular culinary plant is also an excellent perennial for the garden. Low growing and aromatic, thyme makes a wonderful edger for pathways and borders. Thyme can be used as a lawn substitute in well-drained soil because it is tolerant of light foot traffic.

PLANTING

Seeding: Start seeds in cold frame in spring. Many popular hybrids, particularly the ones with variegated leaves, cannot be grown from seed. Common Thyme and Mother of Thyme are good choices to start from seed.

Planting out: Spring.

Spacing: 16" (40 cm) apart.

GROWING

Thyme prefers **full sun**. Soil should be **average or poor and very well drained** and it is beneficial to have leaf mold worked into it. Thyme is a useful plant to include in the front of borders, between or beside paving stones and in rock gardens and walls. Divide plants in spring.

This large genus has species throughout the world that were used in various ways in several cultures. Ancient Egyptians used it in embalming, the Greeks in baths and the Romans to purify their rooms.

T. serpyllum

T. vulgaris

In the Middle Ages, it was believed that to help enable one to see fairies one should drink a thyme infusion.

RECOMMENDED

T. x citriodorus (Lemon-scented Thyme) forms a mound 12" (30 cm) tall and 10" (25 cm) wide. The foliage does smell of lemon. The flowers are pale pink. The cultivars are more ornamental. **'Argenteus'** has silver-edged leaves. **'Aureus'** has golden-yellow leaves.

T. serpyllum (Mother of Thyme; Wild Thyme) is a popular low-growing species. It usually grows about 5" (13 cm) tall and spreads 12" (30 cm) or more. The flowers are purple. There are many cultivars available. **'Minimus'** is a tiny plant growing 2" (5 cm) high and 4" (10 cm) wide. **'Snowdrift'** has white flowers.

T. vulgaris (Common Thyme) forms a bushy mound of dark green leaves. The flowers may be purple, pink or white. It usually grows about 12–18" (30–45 cm) tall and spreads about 16" (40 cm). **'Silver Posie'** is a good cultivar with pale pink flowers and silver-edged leaves.

T. x citriodorus

GARDENING TIPS

Once the plants have finished flowering it is a good idea to shear them back by about half. This encourages new growth and prevents the plants from becoming too woody.

It is easy to propagate the cultivars that cannot be started from seed. As the plant grows outwards the branches touch the ground and send out new roots. These rooted stems may be removed and grown in pots to be planted out the following spring. Unrooted stem cuttings may be taken in early spring, before flowering.

PROBLEMS & PESTS

Thyme rarely has any problems. Seedlings may suffer from damping off and plants may get gray mold or root rot. Good circulation and adequate drainage are good ways to avoid these problems.

For a relaxing bath at the end of a long hard day in the garden, hold a handful of thyme under the running water as you fill the tub and let the leaves float around you as you soak.

T. serpyllum

Toadflax

Linaria purpurea

Flower color: White, purple or pink.
Height: About 36" (90 cm) tall. **Spread:** 12" (30 cm) wide.
Blooms: Early summer to early fall.

These are tall, charming and graceful plants with flowers that resemble snapdragons. Toadflax will bloom even for beginner gardeners who haven't had a chance to improve their rocky, gravel-filled soil.

PLANTING

Seeding: Start seeds in cold frame in early spring.

Planting out: Spring or fall.

Spacing: 12" (30 cm) apart.

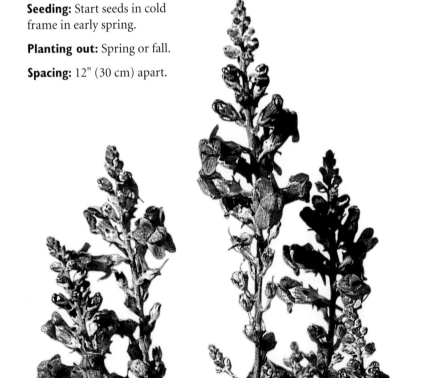

GROWING

Toadflax prefers to grow in **full sun**. Soil should be **fairly fertile, sandy and well drained**. These plants may be used in a border. They are also useful in a dry sandy area where they can be left to naturalize. On a gravely bank they help prevent soil erosion. Divide in spring.

RECOMMENDED

'Canon J. Went' has pale pink flowers and silver gray foliage.

'Springside White' has white flowers.

GARDENING TIPS

You may wish to remove the flower spikes as they fade. These plants self-seed very easily and you may find the population growing a bit too high if you don't do something to prevent it.

PROBLEMS & PESTS

Occasional problems with aphids, flea beetles, downy mildew, white smut and anthracnose are possible.

Make sure you don't plant the related species L. vulgaris, commonly known as Butter and Eggs. This plant, though attractive growing along roadsides and in vacant lots, is an invasive weed that escaped cultivation. It is a problem in native habitats because it doesn't require disturbed land the way many other weeds do.

Trillium
Wake Robin
Trillium spp.

Flower color: White, yellow, pink or red.
Height: 12–18" (30–45 cm) tall. **Spread:** 12" (30 cm) wide.
Blooms: Spring.

These are beautiful woodland plants that require very little care. Be sure to get them from a reputable nursery. Wild populations of these difficult to propagate plants have been severely damaged by people collecting them. Trillium is a welcome sight in early spring.

PLANTING

Seeding: Not recommended. It may take two or more years before you see any growth and another five or more years before plants reach flowering size. Start ripe seeds in shaded cold frame.

Planting out: Fall or spring.

Spacing: 12" (30 cm) apart.

GROWING

Locate plants in **full or partial shade**. Soil should be **humus-rich, moist, well drained and neutral or acidic**. Trillium are ideal plants for woodland gardens or under spring flowering trees and shrubs. Division is not necessary.

RECOMMENDED

T. erectum (Purple Trillium) has deep wine red flowers.

T. grandiflorum (Great White Trillium) has large white flowers that fade to pink. The cultivar **'Flore Pleno'** is slower growing, but has double flowers.

T. ovatum (Coast Trillium) is native to the Pacific coast. The flowers are white and fade to pink or red (photo on opposite page).

GARDENING TIPS

These plants are best planted then left alone. Newly transplanted plants may take a year or two to adjust and start flowering. The amount of moisture they receive after flowering greatly influences how quickly the plants become established. Plentiful moisture in summer prevents the plant from going dormant after flowering. Instead, the plant will send up side shoots that will increase the size of the clump and the number of flowers in spring.

PROBLEMS & PESTS

This plant has few pest problems. Young foliage may be attacked by slugs and snails.

Trillium is a wonderful woodland flower that works well when combined with ferns, violets and epimediums.

T. grandiflorum

Verbena

Verbena spp.

Flower color: Pink, purple or blue.
Height: Upright: 36–72" (90–180 cm) tall. Low-growing: 4–18" (10–45 cm) tall.
Spread: 12–36" (30–90 cm) wide. **Blooms:** Summer.

Colorful verbena laughs at the hot weather, producing even more bright blooms in convivial clusters the hotter it gets. It loves to grow on mounded banks or slight slopes where standing water can't rot its roots. Temperamental when it comes to our cool, wet winters, this is one perennial that is often grown as an annual, replacing it each spring with fresh plants.

PLANTING

Seeding: Direct sow in late summer or fall.

Planting out: Spring or fall.

Spacing: 12–36" (30–90 cm) apart.

GROWING

Grow these plants in a location that receives **full sun**. Soil should be of **average fertility, moist and well drained**. Upright species may be grown in a border and in masses. Low-growing species may be grown on rock walls, at the front of borders or in containers. Verbenas are excellent for filling in the spaces between other plants. Divide in spring or fall every three years or so.

RECOMMENDED

V. bonariensis (Brazilian Vervain) grows up to 72" (180 cm) tall. Light purple flowers are held in clusters at the tops of long stems.

V. bonariensis

V. canadensis (Rose Verbena) is low-growing and can spread up to 36" (90 cm). Flowers are rose pink. This species is grown as an annual (photo on opposite page).

V. hastata (Blue Vervain) is an upright species, growing 36–60" (90–150 cm) tall. The flowers are bluish purple, pinkish purple or white. Blue Vervain thrives in damp ground.

V. tenuisecta (Moss Verbena) can be low growing or somewhat upright. It will grow up to 20" (50 cm) tall, but the stems usually sprawl along the ground. Flowers are blue, white or in shades of purple.

GARDENING TIPS

If the plants begin to look straggly by midsummer they may be sheared back to promote new growth. Mulch in winter to protect this tender perennial or re-plant it each year if it doesn't survive in your garden.

PROBLEMS & PESTS

Verbena may have problems with aphids, whiteflies, slugs, snails, scale insects, spider mites, powdery mildew, leaf spots and rust.

Folklore suggests that historically Verbena was sacred in many cultures. In Egypt, it was believed that Verbena originated from the tears of Isis.

Vinca

Periwinkle; Creeping Myrtle

Vinca minor

Flower color: Blue-purple, sometimes pale blue, reddish purple or white.
Height: 4–8" (10–20 cm) tall. **Spread:** Indefinite.
Blooms: Midspring through to fall.

Vinca is a shade-loving perennial that forms a dense, deep green mat of light veined, evergreen leaves. The first flush of flowers in spring is most plentiful but the plant will continue to produce flowers all summer. This plant is most at home growing in a sun-dappled woodland garden. The flowers are small and five-petalled and are borne singly wherever a leaf joins the stem. Periwinkle grows quickly, but not aggressively, rooting along its stems as it grows.

PLANTING

Seeding: Not recommended.

Planting out: Spring or fall.

Spacing: 24–36" (60–90 cm) apart to get full coverage more quickly.

GROWING

Grow Vinca in **partial to full shade**. It will grow in any type of soil as long as it is **not too dry**. The plants will turn yellow if the soil is too dry or the sun too hot. Vinca is a useful and attractive groundcover under trees or on a shady bank and it will prevent soil erosion. Divide Vinca in early spring or mid- to late fall or whenever it is becoming overgrown. One plant could cover almost any size of area.

RECOMMENDED

'**Argenteovariegata**' has white-margined leaves and light purple-blue flowers.

'**Bowles' White**' or '**Miss Jekyll**' has white flowers that open from pink-tinged buds.

'**Double Burgundy**' has plum purple, double flowers.

'**Ralph Shugart**' has green and white variegated foliage and large blue flowers.

GARDENING TIPS

If Vinca begins to outgrow its space it may be sheared back hard in early spring. The sheared-off ends may root along the stems. These cuttings may be potted up and given away as gifts or may be introduced to new areas of the garden.

This plant also makes an excellent groundcover in a shrub border as it is shallow-rooted and well able to out-compete weeds, but won't interfere with deeper-rooted shrubs.

The glossy green foliage of Vinca is attractive and cooling in the heat of summer, long after the flowering has finished.

The Romans used the long, trailing stems of Vinca to make wreaths. This use of the plant may have given it the name Vinca, which is derived from the Latin word vincire, *meaning 'to bind.'*

Violet

Viola spp.

Flower color: Purple, blue, white, yellow or orange.
Height: 3–8" (8–20 cm) tall. **Spread:** About 16" (40 cm) wide.
Blooms: Spring and summer, sometimes blooming again in fall.

One of the first 'pass along' plants I ever received, the low growing violets soon invaded my new garden, spilling golden seeds in every crack and cranny. I'm still ripping out overly enthusiastic violet volunteers but have learned to love the delicate, early spring blossoms. The flowers embellish the plants as they fill in the empty spaces with lush spring growth around just emerging hosta and winter bare stands of maidenhair ferns.

PLANTING

Seeding: Start freshly ripened seeds in cold frame.

Planting out: Spring.

Spacing: 12" (30 cm) apart.

GROWING

Violets will grow equally well in **full sun or partial shade**. Soil should be **fertile, humus-rich, moist and well drained**. Violets are good for rock gardens, walls and the front of a border but can be invasive. Divide in spring or fall.

RECOMMENDED

V. adunca (Western Dog Violet) is a compact plant growing only 3" (8 cm) tall and wide. The fragrant flowers range in color from purple to blue.

V. cornuta (Horned Violet) is a low-growing, wide-spreading plant. It grows 6" (15 cm) tall and 16" (40 cm) wide. Cultivars are available in all colors.

V. odorata (Sweet Violet) is a sweetly scented plant that grows 8" (20 cm) tall and 12" (30 cm) wide. The flowers are blue or white.

GARDENING TIPS

Violets self-seed freely and you may find them cropping up in unlikely places in the garden. Cultivars may not set seed.

PROBLEMS & PESTS

Slugs and snails may attack these plants. Mildew is a problem if the plants dry out in summer.

V. adunca

In Greek mythology, the violet was the flower of Aphrodite, the goddess of love, and of her son, Priapus, the god of gardens. The Greeks named the violet the symbol of Athens.

The violet is used in holistic as well as conventional medicine.

Wall Rockcress

Arabis caucasica

Flower color: White or pink.
Height: 6" (15 cm) tall. **Spread:** 20" (50 cm) or wider.
Blooms: Mid- to late spring.

Wall Rockcress is a versatile, low-growing, drought-tolerant plant. The small, fragrant flowers are profusely displayed in spring. It quickly fills areas where many plants find the soil too poor to grow. Keep Wall Rockcress in mind when planning your spring garden as it pairs up nicely with spring tulips.

PLANTING

Seeding: Start seeds in cold frame in fall.

Planting out: Any time.

Spacing: 12" (30 cm) apart.

GROWING

Wall Rockcress prefers to grow in **full sun**. Soil should be **average or poor and well drained**. It should have plenty of **lime** mixed in as *Arabis* needs an alkaline soil. This plant will do best in a climate that doesn't have extremely hot weather in summer. It makes a good addition to a rock garden, rock wall or border edging. It may also be used as a groundcover on an exposed slope or as a companion plant in a natural setting with small bulbs. Divide in early fall every two or three years. Stem cuttings, taken from the new growth, may be started in summer.

RECOMMENDED

'Flore Pleno' has white, double flowers.

'Rosabella' has rose pink flowers.

'Spring Charm' has soft rose pink flowers.

'Variegata' has white flowers and white-edged leaves.

GARDENING TIPS

Cutting the plant back after flowering will keep it neat and compact. Wall Rockcress can be an aggressive grower and should not be planted where it may overwhelm slower growing plants.

PROBLEMS & PESTS

White rust, downy mildew and rust are the most common problems. Aphids are occasionally a problem along with *Arabis* midge. *Arabis* midge causes deformed shoots that should be removed and destroyed. Slugs can be a problem in moist soil.

Grow Wall Rockcress in a place where you can enjoy its sweetly-scented flowers.

Arabis *looks quite a lot like* Aubrietia. *They are both commonly known as rockcress, so make sure you write down the botanical name of the plant you want before taking a trip to the garden center to buy it.*

White Gaura

Gaura lindheimeri

Flower color: White, fading to pink.
Height: 60" (150 cm) tall. **Spread:** 36" (90 cm) wide.
Blooms: Late spring or early summer until frost.

Heat-tolerant and sun-loving, White Gaura provides a textural contrast for the thick and rounded leaves of succulents and sedums with its wiry stems of delicate blooms. An easygoing plant, it's not used often enough in Northwest gardens.

PLANTING

Seeding: Start seeds in cold frame in spring or early summer.

Planting out: Spring.

Spacing: 30" (75 cm) apart.

GROWING

White Gaura prefers a location in **full sun**, but will tolerate partial shade. Soil should be **fertile, moist and well drained**. *Gaura* is drought-tolerant once it becomes established. White Gaura is a good addition for borders. Its color and appearance have a softening effect on brighter colors and although it has few flowers at a time it blooms for the entire summer.

'Corrie's Gold'

'Siskiyou Pink'

RECOMMENDED

'Corrie's Gold' has gold variegation around the edges of its leaves.

'Siskiyou Pink' has bright pink flowers.

'Whirling Butterflies' only grows to 36" (90 cm) tall and tends to have more flowers in bloom at a time.

GARDENING TIPS

In order to keep this plant flowering right up until the end of the season it is important to deadhead. Remove the spent flower spikes as they fade.

PROBLEMS & PESTS

Generally pest free, White Gaura may occasionally have problems with rust, fungal leaf spot, downy mildew or powdery mildew.

There are about 20 species of Gaura, and they are all native to North America.

Wild Ginger
Asarum spp.

Flower color: Burgundy and green.
Height: 3–6" (8–15 cm) tall. **Spread:** About 12" (30 cm) or wider.
Blooms: Early summer.

Wild ginger is a beautiful groundcover for woodland sites. Glossy, heart-shaped leaves form a low-growing mat that is quick-growing but not invasive. Several of the available species are North American natives, and two of these natives are from west of the Rockies. This plant is grown for its foliage more than its cup-shaped flowers, which are generally hidden under the leaves.

PLANTING

Seeding: Start seeds in cold frame or direct sow in garden in spring.

Planting out: Spring or fall.

Spacing: 12" (30 cm) apart.

GROWING

Wild ginger needs to be in the **shade or partial shade**. Soil should be **moist and humus-rich**. Wild ginger will tolerate drought for a while in good shade, but prolonged drought will eventually cause the plant to wilt and die back. All *Asarum* species prefer to grow in acid soils, but *A. canadense* will tolerate alkaline conditions. Use wild ginger in a shady rock garden, border or woodland garden. Division is unlikely to be necessary, except to propagate more plants.

RECOMMENDED

A. canadense (Canada Wild Ginger) is native to eastern North American.

A. caudatum (British Columbia Wild Ginger) is native to northwestern North American, preferring cooler summer growing conditions.

A. europaeum (European Wild Ginger) is a European native. The leaves of this plant are very glossy and it is the quickest species to spread over an area (photo on opposite page).

A. hartwegii is native to Oregon and California. The leaves of this plant are dark green with white, mottled veins.

GARDENING TIPS

The thick, fleshy rhizomes grow along the soil or just under it. Pairs of leaves grow up from the rhizomes. Cuttings can be made by cutting off the sections of rhizome with leaves growing from them and planting each section separately. When taking cuttings you must be careful not to damage the tiny thread-like roots that grow from the stem below the point where the leaves grow.

A. caudatum

Wild ginger rhizomes have a distinctive ginger scent to them and though they are not related to the ginger commonly used in cooking they may be used as a flavoring in many dishes.

PROBLEMS & PESTS

Slugs and snails may be problems in spring on the new growth. Leaf galls and rust occur occasionally.

Wild ginger is reputed to have many medicinal benefits. It has been used as a decongestant, a contraceptive and as a digestive stimulant.

Yarrow

Achillea spp.

Flower color: White, yellow, red, orange or pink.
Height: 5–48" (13–120 cm) tall. **Spread:** 15–24" (38–60 cm).
Blooms: Midsummer to early fall.

Yarrow is a popular wildflower turned garden plant. It may be found, in its many forms, growing wild across most of the northern hemisphere. Yarrow has different species adapted for various conditions. Yarrow is renowned for its durability as well as its delicate foliage and flowers. It is used in beds and borders and also as a lawn replacement plant or groundcover. The most common colors in the wild are white and yellow, but garden varieties are available in bright or pastel shades of red, orange and pink.

PLANTING

Seeding: Direct sow on top of soil in spring.

Planting out: Spring.

Spacing: 12–24" (30–60 cm) apart.

GROWING

Grow this plant in **full sun**. Yarrow will tolerate a great deal of abuse, being tolerant of drought and poor soil. Yarrow will do well in an **average to sandy** soil or any soil that is **well drained**. It will not do well in a heavy, wet soil. This plant is not tolerant of high humidity. Yarrow is a very informal plant. It looks best when grown in a natural-looking garden. Cottage gardens and wildflower gardens are perfect places to grow yarrow. It thrives in hot, dry locations where nothing else will grow. Divide every four or five years, in spring or fall.

'Summer Pastels'

A. millefolium

The ancient Druids used yarrow to divine seasonal weather and the ancient Chinese used it to foretell the future.

RECOMMENDED

'**Summerwine**' has dark, rosy red flowers and grows up to 24" (60 cm) tall.

A. filipendulina has yellow flowers and grows up to 48" (120 cm) tall.

A. x *lewisii* '**King Edward**' has pale yellow flowers and grows to about 5" (13 cm) tall.

A. millefolium '**Summer Pastels**' has flowers of all colors and grows up to 24" (60 cm) tall. This is the most heat and drought-tolerant cultivar and has fade-resistant flowers.

A. ptarmica '**Boule de Neige**' has pure white, double flowers and grows 18–24" (45–60 cm) tall.

A. filipendulina

GARDENING TIPS

Yarrow makes an excellent groundcover, despite being quite tall. The plant sends up shoots and flowers from a low basal point and may be mowed periodically without excessive damage to the plant. Mower blades should be kept quite high, i.e., no lower than 4" (10 cm). Keep in mind that you are mowing off the flowerheads. Do not mow more often than once a month, or you will have short yarrow but no flowers!

Remove the flowerheads once they begin to fade. Yarrow will flower more profusely and for a longer period if it is deadheaded.

Yarrow is used in fresh or dried arrangements. Pick flowers only after pollen is visible on the flowers or they will die very quickly.

Yarrow has blood coagulant properties that were recognized by the ancient Greeks. Achillea is named after the legend of Achilles because during the battle of Troy he treated the wounds of his warriors with this herb.

A. millefolium

Quick Reference Chart
HEIGHT LEGEND Low: < 12" (30 cm) • Medium: 12–24" (30–60 cm) • Tall: > 24" (60 cm)

SPECIES by Common Name	COLOR								BLOOMING			HEIGHT		
	White	Pink	Red	Orange	Yellow	Blue	Purple	Foliage	Spring	Summer	Fall	Low	Medium	Tall
Agapanthus	✿					✿	✿			✿	✿			✿
Ajuga	✿	✿				✿	✿	✿	✿	✿		✿		
Artemisia	✿				✿			✿		✿		✿	✿	✿
Astilbe	✿	✿	✿				✿	✿		✿			✿	✿
Baby's Breath	✿	✿								✿			✿	✿
Balloon Flower	✿	✿				✿				✿			✿	
Basket-of-gold					✿				✿			✿		
Bear's Breeches	✿	✿			✿		✿	✿	✿	✿				✿
Bee Balm	✿	✿	✿				✿			✿				✿
Bellflower	✿	✿				✿	✿		✿	✿		✿	✿	✿
Bergenia	✿	✿	✿					✿	✿				✿	
Black-eyed Susan		✿		✿	✿					✿	✿		✿	✿
Bleeding Heart	✿	✿	✿		✿	✿	✿		✿	✿			✿	✿
Blue Star						✿			✿	✿			✿	✿
Candytuft	✿								✿	✿		✿	✿	
Cardinal Flower	✿	✿	✿			✿	✿			✿	✿			✿
Catmint	✿				✿	✿	✿	✿	✿	✿	✿		✿	✿
Cinquefoil	✿	✿	✿	✿	✿					✿		✿	✿	
Clematis	✿	✿	✿			✿	✿	✿		✿	✿		✿	✿
Columbine	✿	✿	✿		✿	✿	✿	✿	✿	✿			✿	✿
Coral Bells	✿	✿	✿					✿	✿	✿		✿	✿	
Coreopsis		✿		✿	✿					✿	✿		✿	✿
Cornflower	✿	✿				✿			✿	✿			✿	
Corydalis	✿		✿		✿	✿		✿	✿	✿	✿	✿	✿	
Cranesbill Geranium	✿	✿	✿				✿	✿		✿		✿	✿	
Cushion Spurge					✿				✿	✿			✿	
Daylily	✿	✿	✿	✿	✿		✿		✿	✿			✿	✿
Delphinium	✿	✿				✿	✿		✿	✿				✿
Dwarf Plumbago						✿				✿	✿	✿	✿	
English Daisy	✿	✿	✿		✿				✿	✿		✿		
Evening Primrose	✿	✿			✿				✿	✿			✿	
False Solomon's Seal	✿							✿	✿					✿
Fleabane	✿	✿		✿	✿		✿			✿		✿	✿	

Quick Reference Chart

Hardy	Semi-hardy	Tender	Sun	Part Sun	Light Shade	Shade	Moist	Well Drained	Dry	Fertile	Average	Poor	Page Number	SPECIES by Common Name
		✿	✿				✿	✿		✿			62	Agapanthus
✿			✿	✿	✿	✿	✿	✿	✿	✿	✿	✿	66	Ajuga
✿	✿		✿					✿	✿		✿	✿	70	Artemisia
	✿			✿	✿		✿	✿		✿			74	Astilbe
	✿		✿					✿			✿		78	Baby's Breath
✿			✿	✿			✿	✿		✿	✿		82	Balloon Flower
✿			✿					✿	✿		✿		84	Basket-of-gold
		✿	✿	✿			✿	✿		✿	✿		88	Bear's Breeches
✿			✿	✿			✿	✿			✿		92	Bee Balm
✿	✿		✿	✿	✿	✿		✿		✿	✿		96	Bellflower
✿			✿	✿			✿	✿	✿	✿	✿		100	Bergenia
✿			✿	✿			✿	✿	✿		✿		104	Black-eyed Susan
✿				✿	✿		✿			✿			106	Bleeding Heart
✿	✿			✿	✿		✿	✿		✿	✿		110	Blue Star
✿	✿		✿				✿	✿			✿	✿	112	Candytuft
✿			✿	✿			✿			✿			114	Cardinal Flower
✿			✿	✿				✿		✿	✿	✿	118	Catmint
✿	✿		✿	✿				✿	✿		✿	✿	122	Cinquefoil
✿	✿		✿	✿			✿	✿		✿			124	Clematis
✿			✿	✿	✿		✿	✿		✿			128	Columbine
✿			✿	✿	✿		✿	✿		✿	✿		132	Coral Bells
✿			✿					✿	✿		✿	✿	134	Coreopsis
✿			✿		✿		✿	✿			✿		138	Cornflower
	✿		✿	✿			✿	✿		✿	✿		140	Corydalis
✿			✿	✿			✿	✿	✿		✿		142	Cranesbill Geranium
✿			✿		✿		✿	✿			✿		144	Cushion Spurge
✿			✿	✿			✿	✿	✿	✿	✿	✿	146	Daylily
✿			✿				✿	✿		✿			150	Delphinium
	✿		✿	✿				✿		✿	✿		154	Dwarf Plumbago
✿	✿		✿	✿				✿		✿	✿		156	English Daisy
✿			✿					✿			✿	✿	158	Evening Primrose
✿					✿	✿	✿				✿		160	False Solomon's Seal
✿	✿		✿		✿		✿	✿	✿	✿	✿		162	Fleabane

Quick Reference Chart

HEIGHT LEGEND Low: < 12" (30 cm) • Medium: 12–24" (30–60 cm) • Tall: > 24" (60 cm)

SPECIES by Common Name	COLOR								BLOOMING			HEIGHT		
	White	Pink	Red	Orange	Yellow	Blue	Purple	Foliage	Spring	Summer	Fall	Low	Medium	Tall
Foamflower	✿	✿						✿	✿			✿	✿	
Forget-me-not	✿	✿			✿	✿			✿	✿		✿		
Foxglove	✿	✿	✿		✿		✿			✿				✿
Gayfeather	✿	✿	✿				✿			✿			✿	✿
Gentian	✿	✿			✿	✿	✿			✿	✿	✿	✿	✿
Geum		✿	✿	✿	✿				✿	✿		✿	✿	
Goat's Beard	✿									✿				✿
Golden Marguerite	✿				✿				✿	✿			✿	✿
Hens and Chicks	✿		✿		✿		✿	✿		✿		✿		
Himalayan Poppy		✿				✿	✿		✿	✿			✿	✿
Hollyhock	✿	✿	✿		✿		✿			✿	✿			✿
Hosta	✿						✿	✿		✿	✿	✿	✿	
Iris	✿	✿	✿		✿	✿	✿		✿	✿		✿	✿	✿
Jacob's Ladder	✿	✿			✿	✿	✿		✿	✿		✿	✿	✿
Japanese Anemone	✿	✿	✿		✿	✿	✿			✿	✿			✿
Japanese Rush								✿				✿	✿	✿
Lady's Mantle					✿			✿		✿	✿		✿	
Lamium	✿	✿	✿				✿	✿	✿	✿		✿		
Lewisia	✿	✿		✿	✿		✿		✿			✿		
Lily-of-the-valley	✿	✿							✿			✿		
Lungwort	✿	✿	✿			✿		✿	✿			✿	✿	
Lupine	✿	✿	✿	✿	✿	✿	✿		✿	✿				✿
Mallow	✿	✿				✿	✿			✿	✿	✿	✿	✿
Marsh Marigold	✿				✿				✿			✿	✿	
Meadow Rue		✿			✿		✿		✿	✿				✿
Meadowsweet	✿	✿							✿	✿			✿	✿
Michaelmas Daisy	✿	✿	✿				✿			✿	✿		✿	✿
Monkshood	✿					✿	✿			✿				✿
Mullein	✿				✿					✿				✿
Oriental Poppy	✿	✿	✿	✿					✿	✿			✿	✿
Pearly Everlasting	✿							✿		✿		✿	✿	✿
Penstemon	✿	✿			✿		✿		✿	✿		✿	✿	
Peony	✿	✿	✿		✿		✿	✿	✿	✿			✿	✿

Quick Reference Chart

Hardy	Semi-hardy	Tender	Sun	Part Sun	Light Shade	Shade	Moist	Well Drained	Dry	Fertile	Average	Poor	Page Number	SPECIES by Common Name
✿				✿	✿	✿	✿			✿	✿		166	Foamflower
✿	✿			✿	✿		✿	✿			✿	✿	168	Forget-me-not
✿				✿	✿		✿	✿	✿	✿			170	Foxglove
✿			✿				✿				✿	✿	174	Gayfeather
	✿	✿	✿	✿			✿	✿			✿		176	Gentian
✿	✿		✿				✿	✿		✿			178	Geum
✿				✿	✿	✿	✿			✿			180	Goat's Beard
✿			✿					✿	✿		✿	✿	184	Golden Marguerite
✿			✿	✿				✿	✿		✿	✿	186	Hens and Chicks
	✿	✿		✿	✿		✿	✿		✿			188	Himalayan Poppy
✿			✿	✿				✿			✿	✿	190	Hollyhock
✿				✿	✿	✿	✿	✿		✿			194	Hosta
✿	✿	✿	✿		✿		✿	✿			✿	✿	198	Iris
✿			✿	✿			✿	✿		✿			202	Jacob's Ladder
✿			✿	✿			✿	✿		✿	✿		204	Japanese Anemone
	✿	✿	✿	✿			✿				✿		206	Japanese Rush
✿			✿		✿		✿			✿			208	Lady's Mantle
✿				✿	✿	✿	✿			✿			210	Lamium
✿	✿		✿					✿	✿		✿		214	Lewisia
✿			✿	✿	✿	✿	✿		✿		✿		216	Lily-of-the-valley
✿				✿	✿	✿	✿			✿			220	Lungwort
✿	✿		✿	✿				✿			✿	✿	222	Lupine
✿	✿		✿	✿			✿	✿			✿	✿	226	Mallow
✿			✿	✿			✿				✿		230	Marsh Marigold
	✿		✿	✿			✿			✿			232	Meadow Rue
✿			✿	✿	✿		✿	✿		✿			236	Meadowsweet
✿			✿	✿			✿	✿		✿			240	Michaelmas Daisy
✿					✿		✿			✿			244	Monkshood
✿	✿		✿					✿				✿	248	Mullein
✿			✿					✿			✿	✿	250	Oriental Poppy
✿			✿				✿	✿			✿	✿	252	Pearly Everlasting
✿	✿		✿	✿				✿	✿	✿			254	Penstemon
✿			✿	✿			✿	✿		✿			256	Peony

Quick Reference Chart

HEIGHT LEGEND Low: < 12" (30 cm) • Medium: 12–24" (30–60 cm) • Tall: > 24" (60 cm)

SPECIES by Common Name	COLOR								BLOOMING			HEIGHT		
	White	Pink	Red	Orange	Yellow	Blue	Purple	Foliage	Spring	Summer	Fall	Low	Medium	Tall
Phlox	✿	✿				✿	✿		✿	✿	✿	✿	✿	✿
Pinks	✿	✿	✿				✿		✿	✿		✿	✿	
Primrose	✿	✿	✿	✿	✿	✿	✿		✿	✿		✿	✿	
Purple Coneflower	✿	✿					✿			✿				✿
Red-hot Poker		✿	✿	✿	✿					✿	✿			
Rock Cress	✿	✿					✿		✿	✿		✿		
Rose Campion	✿	✿					✿			✿			✿	✿
Rose-mallow	✿	✿	✿							✿	✿			✿
Russian Sage						✿	✿			✿	✿			✿
Sandwort	✿							✿	✿	✿		✿		
Saxifrage	✿	✿	✿		✿			✿	✿	✿		✿	✿	
Scabiosa	✿	✿				✿	✿			✿			✿	
Sea Holly	✿					✿	✿			✿	✿		✿	✿
Sea Pink	✿	✿							✿	✿		✿		
Sedum	✿	✿	✿		✿			✿		✿	✿	✿	✿	
Snow-in-summer	✿								✿	✿		✿		
Spike Speedwell	✿	✿				✿	✿			✿			✿	
Stonecress	✿	✿	✿						✿	✿		✿		
Sweet Rocket	✿	✿					✿		✿					✿
Sweet Woodruff	✿							✿	✿	✿		✿		
Thyme	✿	✿					✿	✿	✿	✿		✿		
Toadflax	✿	✿					✿			✿	✿			✿
Trillium	✿	✿	✿	✿					✿			✿	✿	
Verbena		✿					✿			✿		✿	✿	✿
Vinca	✿					✿	✿	✿	✿	✿	✿	✿		
Violet	✿			✿	✿	✿	✿		✿	✿		✿		
Wall Rockcress	✿	✿							✿			✿		
White Gaura	✿	✿							✿	✿	✿			✿
Wild Ginger							✿	✿		✿		✿		
Yarrow	✿	✿	✿	✿	✿			✿		✿	✿	✿	✿	✿

Quick Reference Chart

SPECIES by Common Name

Hardy	Semi-hardy	Tender	Sun	Part Sun	Light Shade	Shade	Moist	Well Drained	Dry	Fertile	Average	Poor	Page Number	Species by Common Name
✿			✿	✿			✿	✿		✿			260	Phlox
✿			✿		✿			✿	✿		✿		264	Pinks
✿	✿		✿	✿	✿	✿	✿	✿		✿	✿		268	Primrose
✿			✿				✿	✿	✿	✿	✿		272	Purple Coneflower
	✿		✿	✿			✿	✿		✿			274	Red-hot Poker
✿	✿		✿	✿				✿			✿		276	Rock Cress
✿			✿	✿				✿			✿		280	Rose Campion
✿	✿		✿				✿	✿		✿			282	Rose-mallow
	✿	✿	✿					✿	✿		✿	✿	286	Russian Sage
✿	✿		✿		✿			✿			✿	✿	288	Sandwort
✿	✿	✿		✿			✿	✿		✿			290	Saxifrage
✿	✿		✿	✿				✿			✿		292	Scabiosa
✿	✿		✿					✿			✿	✿	294	Sea Holly
✿			✿					✿	✿		✿	✿	298	Sea Pink
✿			✿	✿			✿	✿			✿		300	Sedum
✿			✿	✿				✿			✿	✿	304	Snow-in-summer
✿			✿	✿	✿		✿	✿			✿		308	Spike Speedwell
	✿	✿	✿					✿		✿	✿	✿	310	Stonecress
✿			✿	✿			✿	✿			✿	✿	312	Sweet Rocket
✿	✿			✿	✿	✿	✿			✿			314	Sweet Woodruff
✿	✿		✿				✿				✿	✿	316	Thyme
✿	✿		✿					✿	✿	✿	✿		320	Toadflax
✿				✿	✿	✿	✿	✿	✿				322	Trillium
✿	✿		✿				✿	✿			✿		324	Verbena
✿				✿	✿	✿	✿	✿		✿	✿	✿	326	Vinca
✿	✿	✿	✿	✿	✿		✿	✿		✿			328	Violet
✿			✿					✿			✿	✿	330	Wall Rockcress
	✿		✿	✿			✿	✿		✿			332	White Gaura
✿	✿			✿	✿	✿	✿			✿	✿		334	Wild Ginger
✿			✿					✿	✿		✿	✿	336	Yarrow

Glossary

Acid soil: soil with a pH lower than 7.0

Alkaline soil: soil with a pH higher than 7.0

Basal leaves: leaves that form from the crown

Crown: the point at or just below soil level where the shoots join the roots

Cultivar: a plant cultivated for desirable characteristics, such as flower color, leaf variegation or disease resistance, that will pass on these characteristics by seed or vegetative propagation

Damping off: fungal disease causing seedlings to rot at soil level and topple over

Deadhead: to remove spent flowers to maintain attractiveness and encourage a longer blooming period

Disbudding: to remove some flower buds to improve the size or quality of the remaining ones

Dormancy: a period, typically during unfavorable conditions, when a plant ceases growth activity

Double flower: a flower made up entirely of petals, with few or no stamens (see hollyhock, p.190)

Genus: the taxonomic grouping above species level; the first word of the binomial Latin name

Hardy: capable of surviving the adverse climatic conditions of a given region, i.e., rain and cold in the Pacific Northwest

Humus: decomposed, or decomposing, organic material in the soil

Hybrid: a cross breed, either naturally occurring or cultivated, between two or more distinct species within a genus or closely related genera; the hybrid expresses features of each parent plant

Marginal aquatic: a plant that prefers moist or wet conditions; may grow in water up to 12" (30 cm) deep; common around ponds or in bog gardens

Neutral soil: soil with a pH of 7.0

Node: the area on a stem from which a leaf or new shoot grows

Offset: the young plants that naturally sprout around the base of the parent plant

pH: the scale used to measure acidity or alkalinity; the pH of soil influences the availability of nutrients

Rhizome: a food-storing stem that grows horizontally at or just below soil level, from which new shoots may emerge

Root ball: a mass of soil and plant roots

Semi-hardy: a plant capable of surviving the climatic conditions of a given region if protected, i.e., from excessive rain in the Pacific Northwest or from the cold in the mountains and interior

Semi-double flower: a flower with petals that form two or three rings (see marsh marigold, p. 230, or peony, p. 256)

Single flower: a flower with petals that form a single ring, typically of four or five petals (see cinquefoil, p. 122)

Species: a group of plants that can interbreed with each other, but that have distinct individual characteristics

Straight species: the original species from which cultivars and varieties are derived

Taproot: a root system consisting of one main root with smaller roots branching from it

Tender: incapable of surviving the climatic conditions of a given region and requiring shelter, i.e., from heavy rain or excessive cold in the Pacific Northwest

True: when desirable characteristics are passed on from the parent plant to the seed-grown offspring; also called breeding true to type

Tuber: the thick section of a rhizome bearing nodes and buds

Variety: a naturally occurring variation or subdivision

Index

W